# MATH + − THE EASY WAY

Your Key to Learning

# MATH × ÷ THE EASY WAY

**Your Key to Learning**

By
**Anthony Prindle**
Chairman, Mathematics Department
Linton High School, Schenectady, New York

Edited by
**Eugene J. Farley**
Consultant in Education

BARRON'S EDUCATIONAL SERIES, INC.
Woodbury, New York • London • Toronto • Sydney

*All inquiries should be addressed to:*
Barron's Educational Series, Inc.
113 Crossways Park Drive
Woodbury, New York 11797

*Library of Congress Catalog Card No. 81-20665*

International Standard Book No. 0-8120-2503-2

**Library of Congress Cataloging in Publication Data**

Prindle, Anthony.
Math the easy way.

1. Mathematics — 1961- I. Title.
QA39.2.P7 513'.14 81-20665
ISBN 0-8120-2503-2 AACR2

PRINTED IN THE UNITED STATES OF AMERICA

567 800 17 16 15 14 13 12 11 10

# Contents

# Introduction

To survive in our world today requires two skills: the ability to read well and a knowledge of mathematics. Without a mastery of mathematics, you find yourself handicapped not only in coping with your job but also in dealing with your personal finances. The purpose of this book is to refresh skills you once learned and to develop new ones.

Many people have a fear of mathematics, just as we all fear the unknown. With a little determination and some practice on your part, however, the secrets of this subject will be revealed and you will find yourself comfortable in dealing with the ideas of mathematics.

## YOUR KNOWLEDGE

Although you forget some details or facts that were learned, you also retain a tremendous amount of information. School classes are a good way to learn, but work experience, family living, newspapers, radio, television, and contact with other people are also useful. It is even more important to think about and use the knowledge. You can probably think of something in your personal or work life that you do very well. In the beginning you had to learn or store the necessary information. Now you do well because you *use* that information rather than let it rest in storage.

## HOW TO STUDY

You must develop a way to study. You decide the time, the place, and the subjects you must practice. However, certain basic rules of study are helpful.

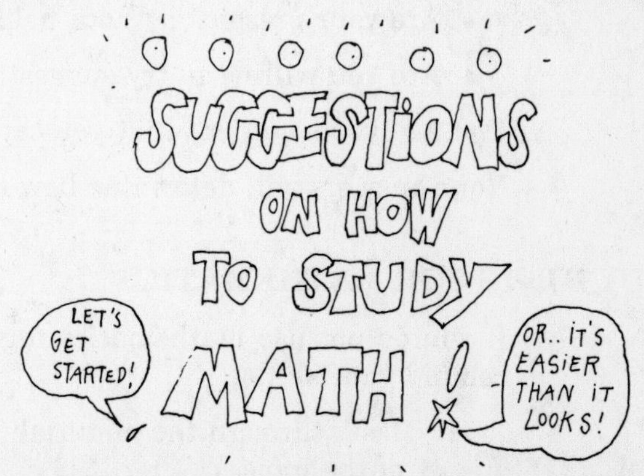


1. *Start with a goal.* Study because you want to learn. As an adult you make your own choice that study is important to your progress.
2. *Make the conditions help you, not distract you.* The place, the furniture, the lighting, and the temperature should not interfere with study.

3. *Organize your materials.* Usually these are the book you are using, the pencils or pens to make notes, and paper or a notebook. Get these ready before you start so that you do not waste your time looking for each item.
4. *Begin.* The greatest enemy of study is the tendency to delay or postpone. Try to set aside a definite time each day so that you are used to studying then.
5. *Survey.* Look over quickly what you plan to do. Get a general idea so that your efforts are directed to that goal.
6. *Take time to think.* Relate the ideas you learn to your previous knowledge.
7. *Make notes.* By writing down the key ideas or main rules you help yourself to learn them. The notes are useful for review at a later time. You can also use notes to list comments or questions which are not clear so that you remember to ask for help or explanation.
8. *Be reasonable in your study time.* Most human beings cannot work long hours without an occasional break. On the other hand long breaks and short study time will cover very little.
9. *Be fair in rating yourself.* You cannot expect to be perfect. Give yourself a chance to improve or make progress.
10. *Review.* Ideas slip away if you do not refer to them again. Review by reading your notes, looking at main headings or asking yourself questions about the material.
11. *Use what you learn.* Practice your mathematical skills in shopping, purchasing, banking, or working.

We have tried to digest books and articles about *how to study* into a few brief rules. If you have a technique that works better for you than one or two of these rules, then you should use it. If you find even these few rules seem like too many to follow, you have to ask yourself the following questions:

- Are your present methods helping you to learn and to reach your goal?
- Are you willing to try suggestions that have worked for many others?
- Are you confident that you can improve by making an effort?

Your answers will determine how much you need the study rules.

## SPECIAL STUDY HINTS FOR MATHEMATICS

If you do not use mathematics regularly these suggestions may help you to rebuild your skills:

1. Look through the material (a section or chapter) to get a general idea of the topic.
2. Note especially any new terms or symbols and their explanations.
3. Read the instructions and explanations before working any problem.

4. Concentrate on *why* (the reasoning which governs the problem) and *how* (the method or process to solve the problem).
5. Follow the order or sequence of steps given in the examples.
6. Consider why you did each step. (In other words, you should be able to explain your reason for doing the work that way.)
7. Use practice problems to check whether you have learned the reasoning (why) and the method (how) for that type of problem.
8. Recognize that mistakes may be frequent at first. You have to persevere and try to hold your frustration in check. Use your mistake to go back through the steps to see how the error developed.

## READING IN MATHEMATICS

Develop the habit of careful reading in mathematics. Do not be alarmed if your reading rate is slow — it should be. Often you must reread a mathematical problem to be sure that you understand. As you read carefully you will be looking for the answers to these questions:

1. What is given? (the facts in the problem)
2. What is unknown? (the answer to be found)
3. How do I proceed? (the methods or steps to solve the problem)

This is a different style of reading from the one you might use in looking through a newspaper or magazine. In mathematics directions and problems are compressed into very few words. The use of symbols and numerals further reduces the number of words. Therefore each word or symbol becomes very important and should not be missed by speedy reading.

As in any subject, your reading in mathematics improves as you learn more terms and symbols. Further, the reading of mathematics requires close attention to relationships. How does one fact or idea lead to another? Which facts or ideas are connected? At first these suggestions may seem frightening to some people. The book will start with easy ideas, use your experience whenever possible, and explain new ideas before expecting you to work with them. Pay close attention to the examples throughout the text. These will show you how to set up a problem, work it out, and check your answer. You know you can do it on your own if we show the way and you practice the skills.

# Chapter 1

# WHOLE NUMBERS

# 1.1 Laws and Operations

All of us are familiar with whole numbers. They are the numbers used in counting: 0, 1, 2, 3, 4 and so on. We will begin with a brief review of operations on whole numbers. An *operation* in mathematics is a way of combining two or more numbers. There are two basic operations that can be performed on numbers: *addition* and *multiplication*. Addition is concerned with totaling up *individual* numbers, as in $3 + 4$, while multiplication combines *groups* of numbers. Five times seven ($5 \times 7$) means five groups of seven or seven groups of five.

When one number is added to another, the answer is the same regardless of the order in which the numbers are added: both $3 + 4$ and $4 + 3$ equal 7. This fact is known as:

| THE COMMUTATIVE LAW OF ADDITION |
|---|

Similarly, you get the same answer in *multiplying* two numbers regardless of the order in which you multiply them: both $5 \times 7$ and $7 \times 5$ equal 35. This is called:

| THE COMMUTATIVE LAW OF MULTIPLICATION |
|---|

You may have questioned the statement above that there are only two basic operations which we perform on numbers. You may feel that there are four fundamental operations: addition, subtraction, multiplication, and division. However, subtraction and division are really opposite or *inverse* operations of addition and multiplication, respectively. In other words, subtraction must be performed to undo an addition, and division must be performed to undo a multiplication.

*Example:* A number plus 5 is equal to 8. What is the number?

The word "plus" suggests that an addition is involved. The number is 3, since $3 + 5 = 8$. But how was 3 obtained as an answer? Since it is clear that $3 = 8 - 5$, the subtraction was performed to *undo* the addition.

In the same way, a division is performed to *undo* a multiplication.

*Example:* A number times 5 is equal to 40. What is the number?

The word "times" suggests that a multiplication is involved. The number is 8, since $5 \times 8 = 40$. We derived this answer by dividing 40 by 5 ($40 \div 5 = 8$). In essence, the division was performed to *undo* the multiplication.

Also, when adding a column of figures, you know that it is permissible to take the numbers in any order. Many people group them in different ways.

Let's examine two different ways of finding the sum of $3 + 5 + 8$:

| | |
|---|---|
| $(3 + 5) + 8$ | $3 + (5 + 8)$ |
| $(8) + 8$ | $3 + (13)$ |
| $16$ | $16$ |

We see again that the grouping of the numbers to be added makes no difference in the result obtained. Similarly, let us examine two different ways of finding the product of $3 \times 5 \times 8$:

| | |
|---|---|
| $(3 \times 5) \times 8$ | $3 \times (5 \times 8)$ |
| $(15) \times 8$ | $3 \times (40)$ |
| $120$ | $120$ |

The fact that we get the same sum when adding three or more numbers, regardless of which way we group them, is called:

THE ASSOCIATIVE LAW OF ADDITION

Likewise, the fact that we obtain the same product, regardless of how we group the numbers to be multiplied, is called:

THE ASSOCIATIVE LAW OF MULTIPLICATION

Multiplication and addition can be linked together, as illustrated in both versions of the following example:

| | |
|---|---|
| $3 \times (5 + 4)$ | $3 \times (5 + 4)$ |
| $3 \times (9)$ | $(3 \times 5) + (3 \times 4)$ |
| $27$ | $15 + 12$ |
| | $27$ |

This final law of operation is called:

THE DISTRIBUTIVE LAW

In the second version above, we have "distributed" the 3 over the two addends, 5 and 4, which make up the sum enclosed in parentheses. Although you may not have used the term *distributive*, it's likely that you have used the law whenever you have multiplied a two-digit number by a single-digit number. For example, since 27 is really $20 + 7$, $5 \times 27$ can be computed as $5 \times (20 + 7)$, or $100 + 35 = 135$.

Later, in Chapters 5–7 on algebra, you will use letters in place of numbers, and these letters will be called *variables*. Now let's use the letters *a, b,* and *c* to state these three basic laws. In these examples, the variables *a, b,* and *c* represent any number.

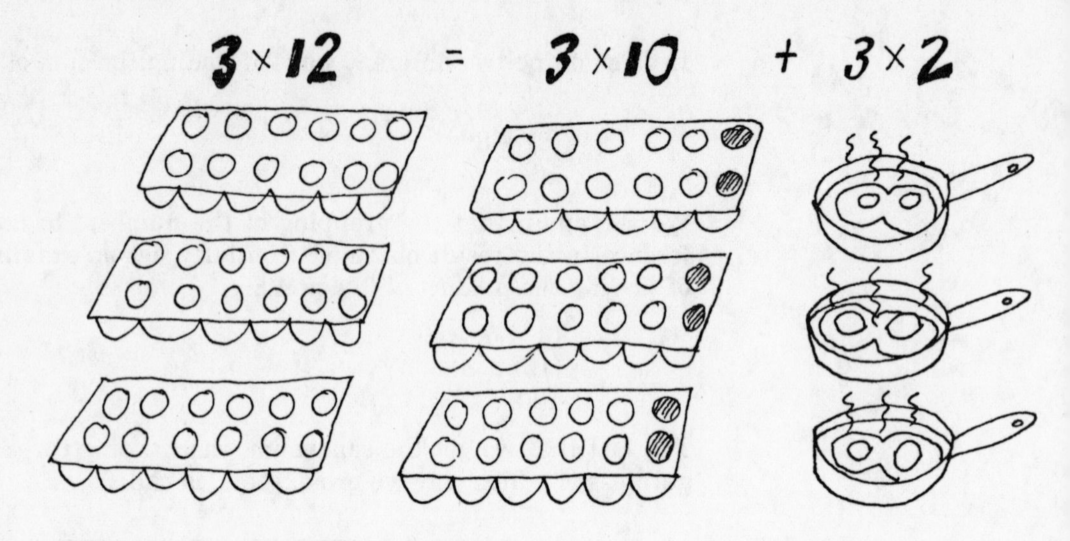


**Commutative Laws:**

| | |
|---|---|
| For *addition:* | $a + b = b + a$ |
| For *multiplication:* | $a \times b = b \times a$ |

**Associative Laws:**

| | |
|---|---|
| For *addition:* | $a + (b + c) = (a + b) + c$ |
| For *multiplication:* | $a \times (b \times c) = (a \times b) \times c$ |

**Distributive Law:**

| | |
|---|---|
| Also called the *distribution of multiplication over addition* | $a \times (b + c) = (a \times b) + (a \times c)$ |

## WHEN NOT TO USE THESE LAWS

Do these same laws apply to the inverse operations of subtraction and division? Is it true that $5 - 3 = 3 - 5$, or $5 \div 3 = 3 \div 5$? Certainly not. $5 - 3$ is equal to 2, but what is $3 - 5$? It means "subtract 5 from 3" and for just a while we'll have to say that this is not possible. Similarly, $5 \div 3$ is equal to $\frac{5}{3}$ or $1\frac{2}{3}$, while $3 \div 5$ is simply $\frac{3}{5}$. So we say that the operations of subtraction and division are *not* commutative.

Now let's test the associative law:

$$5 - (3 - 1) = 5 - 2 \text{ or } 3,$$
$$\text{while } (5 - 3) - 1 = 2 - 1 \text{ or } 1.$$

Clearly, subtraction is *not* associative.

$$18 \div (6 \div 3) = 18 \div 2 \text{ or } 9,$$
$$\text{while } (18 \div 6) \div 3 = 3 \div 3 \text{ or } 1.$$

Similarly, division is *not* associative.

**Exercises**

Try to identify the law of operation that is used in each of the following:

1. $(4 + 7) + 12 = 4 + (7 + 12)$
2. $3 \times 8 = 8 \times 3$
3. $7 \times 25 = 7 \times 20 + 7 \times 5$

**Answers**

1. Associative law for addition.
2. Commutative law for multiplication.
3. Distributive law [because $7 \times 25$ is really $7 \times (20 + 5)$].

## ORDER OF OPERATIONS

The order in which operations are performed is very important. We must be able to agree on the answer to the problem $2 + 3 \times 5$ Some people might add 2 and 3 first, thus getting $5 \times 5$ or 25. Others might multiply $3 \times 5$ first, thus getting $2 + 15$ or 17. Is it 25 or 17? Mathematicians have agreed that, in any problem:

1. Multiplications and divisions are to be carried out *before* additions and subtractions.
2. When *several* multiplications and divisions occur together, you should carry out multiplications and divisions *in the order they occur*.

   *For example:* $12 \div 6 \times 2 = 4$, *not* 1.

Try these exercises to make sure you understand the order of operations.

**Exercises**

1. $3 + 8 \times 7$
2. $12 + 4 \div 2$
3. $28 - 14 \div 7 + 4$
4. $18 \div 2 + 4 \times 3 - 21$
5. $32 - 16 \times 2 + 5$

**Answers**

1. 59 2. 14 3. 30 4. 0 5. 5

# 1.2 Factorization and Divisibility

Before we start the discussion of this topic, there are a few key words you ought to know. Your ability to solve problems is no better than your understanding of the words used.

DEFINITIONS

**Sum** is the answer in addition.

**Difference** is the answer in subtraction.

**Product** is the answer in multiplication.

**Quotient** is the answer in division.

The whole numbers 0, 1, 2, 3, . . . are divided into two types: primes and composites. To help you understand these two kinds of numbers we now introduce two words: *factor* and *divisor.*

DEFINITIONS

**Factors** are multipliers used to form a product.
*Examples:* 3 and 7 are *factors* of 21. 1 and 21 are *factors* of 21.

**Divisor** has the same meaning as factor, except that it is used in division.
*Example:* 3 is a *divisor* of 21 (21 is divisible by 3).

Since $3 \times 7 = 21$, we say that 3 and 7 are *factors* of 21. Likewise, 1 and 21 are *factors* of 21 because $1 \times 21 = 21$. The four numbers 1, 3, 7, and 21 are all factors of 21.

A *divisor* is the same as a factor when we say "3 is a *divisor* of 21" or "21 is divisible by 3." In other words, when 21 is divided by 3, there is no remainder.

Now suppose 21 is divided by 4. There is a remainder of 1, so 4 is not a divisor of 21, and 21 is not divisible by 4.

With these two new words, we can now make good definitions for two other key words, *prime* and *composite*.

DEFINITIONS

A **prime number** is a whole number whose only divisors are itself and 1.
*Examples:* 2, 3, 5, 7, 11, 13

A **composite number** is one which has at least one other factor beside itself and 1.
*Examples:* 4, 6, 8, 9, 10, 12

NOTE: *Every whole number must be either prime or composite, with two exceptions, 0 and 1. These are neither prime nor composite.*

Think again about the word *factor*. If you are asked to *factor* a number such as 24, this means that you are to find multipliers which will have a product of 24. You might think that the answer could be any of the following pairs:

| | | |
|---|---|---|
| $1 \times 24$ | Using the commutative law, you may write the factors in reverse order. | $24 \times 1$ |
| $2 \times 12$ | | $12 \times 2$ |
| $3 \times 8$ | | $8 \times 3$ |
| $4 \times 6$ | | $6 \times 4$ |

Actually none of these is the factorization we want, for when we say "factor," we agree that we mean only *prime factors*. The prime factorization of 24 is $2 \times 2 \times 2 \times 3$, and the prime factors of 24 are the numbers 2 and 3.

Before you attempt to factor any numbers, it will be advisable for you to memorize the ten smallest primes: 2, 3, 5, 7, 11, 13, 17, 19, 23, 29. Then, too, it will help you in factoring to review the following ways in which you can determine, without actually dividing, whether the number you wish to factor is divisible by certain numbers:

FACTORING HINTS

| **Clues** | **Divisor** |
|---|---|
| Is the unit's digit 0, 2, 4, 6, or 8? ............................ | **2** |
| Is the *sum* of all the digits divisible by 3? ..................... | **3** |
| Is the unit's digit 0 or 5? .................................... | **5** |
| Is the *sum* of all the digits divisible by 9? ..................... | **9** |
| Is the unit's digit 0? ........................................ | **10** |

*For example:* 12, 284 and 3,178 are all divisible by **2**, because the unit's digit is an even number.

36, 279 and 1,536 are all divisible by **3** because in each case the sum of the digits is divisible by 3 (9, 18 and 15 are divisible by 3).

75, 120 and 1,875 are divisible by **5**.

126, 81 and 729 are divisible by **9**, since the sum of the digits (9, 9 and 18) is divisible by **9**.

120, 50, and 1,970 are all divisible by **10** since the final digit is 0.

## HOW TO FIND THE PRIME FACTORS OF A NUMBER

STEP 1 Find the smallest prime factor of that number and divide by it to obtain a second factor:

$$2\overline{)24}^{\;12}$$

STEP 2 If the second factor is not prime, find the smallest prime that will divide it; and divide by that prime to obtain a third factor:

$$\begin{array}{r}12\\2\overline{)24}\end{array},\quad \begin{array}{r}6\\2\overline{)12}\end{array}$$

STEP 3 If the third factor is not prime, repeat the process until the final factor is prime:

$$\begin{array}{r}12\\2\overline{)24}\end{array},\quad \begin{array}{r}6\\2\overline{)12}\end{array},\quad \begin{array}{r}3\\2\overline{)6}\end{array}$$

The prime factorization of 24 is:

$$24 = 2 \times 2 \times 2 \times 3$$

**Exercises** Determine which numbers are prime (divisible *only* by the number itself and 1), and which are composite. If the number is composite, find its prime factorization.

NOTE: You will find it helpful to try the prime divisors *in order* from your list of primes. This will tend to keep you from omitting factors.

**1.** 37 **2.** 49 **3.** 125 **4.** 63 **5.** 61

**6.** 119 **7.** 105 **8.** 147 **9.** 151 **10.** 289

**Answers**

**1.** prime **2.** $7 \times 7$ **3.** $5 \times 5 \times 5$ **4.** $3 \times 3 \times 7$

**5.** prime **6.** $7 \times 17$ **7.** $3 \times 5 \times 7$ **8.** $3 \times 7 \times 7$

**9.** prime **10.** $17 \times 17$

You will observe that it is customary in writing answers to list factors in order of size as given above. However, since multiplication is commutative, any order is acceptable.

## WORKING WITH EXPONENTS

There is a short way of writing repeated factors in multiplication. For example, we may write $5 \times 5$ as $5^2$. The small 2 written to the right and slightly above the 5 is called the exponent. It tells us that the 5 is used twice as a factor. $5^2$ is read either "5 to the *second power*," or more briefly "5 *squared*." *Note that it does not represent* $5 \times 2$. Similarly: $2 \times 2 \times 2 \times 5 \times 5$ written with exponents or in exponential form becomes $2^3 \times 5^2$. The "3" is the exponent of 2 and the "2" is the exponent of 5. The expression may be read "2 *cubed* times 5 *squared*." You may also read it: "2 to the third power times 5 to the second power." The exponents 2 and 3 are the only ones which have special names, "square" and "cube" respectively. For example $5^4$ is read: "5 to the fourth power," or simply "5 to the fourth."

*Example:* Write the prime factorization of 288 in *exponential form*.

```
2 | 288
2 | 144
  2 | 72
  2 | 36
  2 | 18
    3 | 9
        3 (prime)
```

$288 = 2 \times 2 \times 2 \times 2 \times 2 \times 3 \times 3 = 2^5 \times 3^2$.

This is read "2 to the fifth times 3 squared."

# 1.3 Rounding Off

Suppose a used car costs $875. You might say that in "round numbers to the nearest hundred" it costs $900. "Rounding off," "estimating" and "approximating" all mean about the same thing. You make a guess as to the *approximate value*. The number scale below will help us to visualize exactly what we are doing in rounding off.

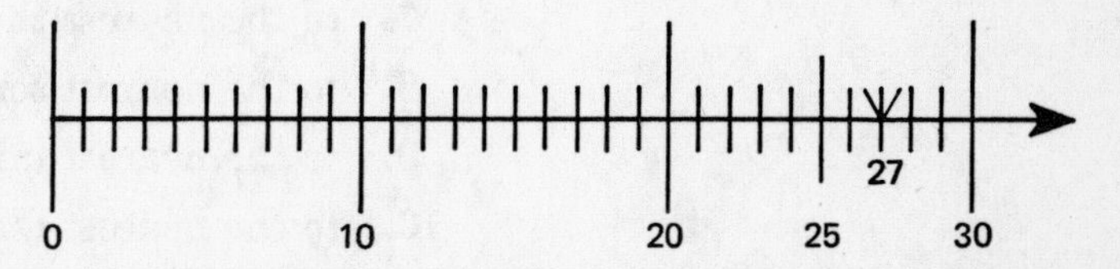


Suppose you wish to round off 27 to the nearest ten. Clearly 27 lies between 2 tens (20) and 3 tens (30). The number 25 lies precisely in the middle between 20 and 30. Since 27 lies 2 spaces to the right of 25, it is closer to 30. Hence you say "27 rounded to the nearest 10 is 30."

How would you apply this reasoning to the number 25? Clearly 25 is precisely in the middle between 20 and 30. It, therefore, is not closer to one than it is to the other. It has been decided arbitrarily that when a number's place on the scale is exactly in the middle, you round it to the next higher. So 25 rounded to the nearest ten also is 30.

Think about the car once again. Round off the price to the nearest thousand dollars. $875 lies between $0 and $1,000. Is $875 closer to $0 or $1,000? A little thinking will convince you that $875 is far closer to $1,000 than to $0. Had you wished to round the same number to the nearest ten, you would have seen that the number was precisely in the middle between 87 tens and 88 tens. From our decision in the preceding paragraph, you would say that $875 rounded to the nearest ten would become $880.

In rounding off a number of any degree of precision, you will find it helpful to follow the procedure outlined below.

*Examples:* **1.** Round off 72,543 to the nearest thousand.

The digit 2 is in the *thousand's* place. The digit to the right of 2 is 5, showing that you are *at least* halfway to the next thousand. Hence, 72,543 becomes 73,000 when rounded to the nearest thousand.

2. Round off 193,$\underline{4}$78,655 to the nearest million.

The digit 3 is in the *million's* place. The digit to the right of 3 is 4, showing we are *less* than halfway between 193 million and 194 million. So, 193,478,655 becomes 193,000,000 when rounded to the nearest million.

**Exercises**

Round off each of the following to the indicated degree of precision:

1. 2,137 to the nearest ten
2. 41,456 to the nearest thousand
3. 32,687 to the nearest hundred
4. 87,524,671 to the nearest million
5. 99,500 to the nearest thousand

Round off 123,546,789 to the indicated degree of precision:

6. to the nearest hundred
7. to the nearest thousand
8. to the nearest ten thousand
9. to the nearest million
10. to the nearest ten million

**Answers**

**1.** 2,140 **2.** 41,000 **3.** 32,700
**4.** 88,000,000 **5.** 100,000 **6.** 123,546,800
**7.** 123,547,000 **8.** 123,550,000 **9.** 124,000,000
**10.** 120,000,000

# Trial Test — Whole Numbers

These questions are arranged in order of difficulty, beginning with the simplest and progressing to the more difficult. If you score 8 or more correct, go on to the next section. If not, read the instructional material once more to clear up difficulties, and try again.

All the answer choices are numbered, so be careful not to confuse an answer with the number of the proper choice. For example, if the correct answer to a problem is 5, and this is answer choice (3), make sure that you mark the third space on the answer grid, *not* the fifth.

1. An example of a prime number is
(1) 27
(2) 31
(3) 52
(4) 1
(5) 91

**1.** 1 2 3 4 5
|| || || || ||

2. Which of the following is an example of the commutative law for addition?
(1) $(3 + 2) + 4 = 3 + (2 + 4)$
(2) $3 \times (2 + 5) = 3 \times 2 + 3 \times 5$
(3) $(3 \times 2) \times 5 = 3 \times (2 \times 5)$
(4) $8 + 5 = 5 + 8$
(5) $3 \times 5 = 5 \times 3$

**2.** 1 2 3 4 5
|| || || || ||

**3.** The number 143 is divisible by
(1) 3 (4) 14
(2) 5 (5) 7
(3) 11

3. 1 2 3 4 5

**4.** The prime factorization of 28 is
(1) $1 \times 28$ (4) $1 \times 7 \times 4$
(2) $2 \times 14$ (5) $2 \times 2 \times 7$
(3) $4 \times 7$

4. 1 2 3 4 5

**5.** If $N = 12 - 4 \times 2 + 3$, then the value of $N$ is
(1) 7 (4) 26
(2) 40 (5) 1
(3) 19

5. 1 2 3 4 5

**6.** To compute $7 \times 43$, one law that must be used is the
(1) commutative law for addition
(2) associative law for addition
(3) commutative law for multiplication
(4) associative law for multiplication
(5) distributive law

6. 1 2 3 4 5

**7.** A number which is divisible by both 5 and 2 is
(1) 3,155 (4) 3,110
(2) 4,212 (5) 4,486
(3) 6,285

7. 1 2 3 4 5

**8.** The difference between the sum of 26 and 14, and the quotient of 12 and 4 is
(1) 7 (4) 8
(2) 37 (5) 41
(3) 4

8. 1 2 3 4 5

**9.** Round off 86,471 to the nearest thousand.
(1) 86 (4) 87,000
(2) 87 (5) 86,500
(3) 86,000

9. 1 2 3 4 5

**10.** A common factor for two numbers is one which is a divisor of each of the two numbers. What is the greatest common factor of 24 and 36?
(1) 12 (4) 4
(2) 2 (5) 6
(3) 3

10. 1 2 3 4 5

## SOLUTIONS

**1.** **(2)** 31 is the only prime number. $27 = 3 \times 3 \times 3$, $52 = 2 \times 2 \times 13$, 1 is neither prime nor composite, and $91 = 7 \times 13$.

**2.** **(4)** The commutative law for addition states that the order in which we add has no effect upon the sum.

3. **(3)** 143 ÷ 11 = 13 with 0 as a remainder. Each of the other numbers used as a divisor will result in a remainder *other* than 0.

4. **(5)** 2 × 2 × 7 is the prime factorization. Each factor is a prime, and the product is 28.

5. **(1)** The law of order of operations states that multiplication and division must be performed *before* addition and subtraction.
   12 − 4 × 2 + 3 = 12 − 8 + 3 = 4 + 3 = 7

6. **(5)** Distributive law. 7 × 43 = 7 × (40 + 3) or 7 × 40 + 7 × 3.

7. **(4)** 3,110. In order that a number be divisible by 5 and 2, it must be divisible by their product, 10. Any number whose unit's digit is zero is divisible by both 2 and 5, and therefore also by 10.

8. **(2)** 37. The use of the words "difference," "sum," and "quotient" furnishes the clues. The sum of 26 and 14 is 40; the quotient of 12 divided by 4 is 3; and the difference is 40 − 3 or 37.

9. **(3)** 86,000. Two frequent errors are shown in choice (1): 86, instead of 86,000, and choice (4): 87,000, obtained by rounding upward rather than downward.

10. **(1)** 12. 2, 3, 4, 6, and 12 are all divisors of both 24 and 36. Of these 12 is the greatest.

## EVALUATION CHART

Multiply the number of correct answers by 100 and divide by the number of questions in the Trial Test to arrive at your score. ____________

| Score | Rating |
|---|---|
| 90–100 | Excellent |
| 80–89 | Very Good |
| 70–79 | Average |
| 60–69 | Passing |
| 59 or Below | Very Poor |

# Chapter 2

# FRACTIONS

Before beginning the review of fractions, let us examine, in a little more detail, our two basic operations, addition and multiplication, as well as the two opposite or *inverse* operations subtraction and division.

# 2.1 Inverse Operations

The two operations: "adding ten" and "subtracting ten" have the effect of "erasing" each other or of *undoing* each other.

*Example:* Think about the number 15. If we first add 10 to it and then subtract 10 from the result, what happens? Right! We're back where we started, at 15.

$$15 + 10 - 10 = 25 - 10 = 15$$

> For each whole number, the operations of *adding* and *subtracting* that number are **inverses** of each other.

Now let's try the operations of multiplying by a number and then dividing by the same number.

*Example:* $50 \times 10 \div 10 = 500 \div 10 = 50$
$50 \div 10 \times 10 = 5 \times 10 \quad = 50$

In each case you have performed two operations, but the labor has left the original number unchanged. Each operation is the opposite of the other.

> For each whole number (with the exception of zero), the operations of *multiplying* and *dividing* by that number are **inverses** of each other.

Exercises

As before, each of the letters $a$, $b$, $c$, and $n$ represents any whole number.

**1.** $813 - 7 + 7$

**2.** $52 + 10 - 10$

**3.** $5 + n - 5$

**4.** $17 - n + n$

**5.** $c + n - n$

**6.** $132 \div 2 \times 2$

**7.** $10 \times 10 \div 10$

**8.** $a \div 5 \times 5$

**9.** $(a + b) \div 3 \times 3$

**10.** $(7 + 10 - 7) \div 5 \times 5$

Answers **1.** 813 **2.** 52 **3.** $n$ **4.** 17 **5.** $c$ **6.** 132 **7.** 10 **8.** $a$ **9.** $a + b$ **10.** 10

The operations of subtraction and division are called *inverse operations* because they are really the *opposites* of addition and multiplication, respectively.

Think about a simple problem:

$$5 - 2 = 3$$

This is really the same as saying $5 = 2 + 3$. Do you see now that subtraction and addition are inverse to each other?

In the same way:

$$12 \div 3 = 4$$

This is really the same as saying that $12 = 3 \times 4$, showing that division ana multiplication are inverse to each other.

In order to make these two definitions of the inverse operations of subtrac tion and division more meaningful, it is necessary for us to examine some special properties of two whole numbers, namely *zero* and *one*.

*Zero has special properties in addition and multiplication:*

THE ADDITIVE PROPERTY OF ZERO

You know that $5 + 0 = 5$ and $0 + 12 = 12$, or more generally: no matter what number $a$ represents, $a + 0 = a$. It is clear that when zero is added to any number, the number is not changed. For this reason *zero is called the identity number for addition.*

THE MULTIPLICATIVE PROPERTY OF ZERO

You know, too, that $3 \times 0 = 0$ and $115 \times 0 = 0$ and $0 \times 0 = 0$, or more generally: no matter what number $a$ represents, $a \times 0 = 0$ and $0 \times a = 0$. This property of zero causes some problems when we try to *divide* by zero. Let us look at the problem:

$$5 \div 0$$

Is the quotient 5? Let us apply our test to prove that 5 *is* or *is not* the quotient. Maybe you think that $5 \div 0 = 5$. If this is really true, then it must be true that $5 = 0 \times 5$. But that cannot be true, because $0 \times 5 = 0$. So you see that *whenever any number is divided by zero, there just is no answer.*

REMEMBER: Never divide by zero.

*One has a special property in multiplication and division.*

Just as adding zero to any number leaves that number unchanged, so does multiplying any number by one leave that number unchanged: $5 \times 1 = 5$. *Thus one is the identity number for multiplication.* Since $5 \times 1 = 5$, it follows that $5 \div 5 = 1$. *Any number, except zero, divided by itself equals one.*

> NOTE: Remember this exception, for $0 \div 0$ does not represent a specific number. Remember too that *you can divide zero, but you cannot divide any number by zero.*

Here is a summary of the last section:

$$c - b = a \text{ if } a + b = c$$

$$c \div b = a \text{ if } a \times b = c$$
*and* $b \neq 0$
($b$ does not equal 0)

PROPERTIES OF ZERO AND ONE

$$a + 0 = a$$
$$a \times 0 = 0$$
$$a \times 1 = a$$
$$a \div a = 1$$
where $a \neq 0$

The expressions $a \div 0$ and $0 \div 0$ do not represent numbers.

# 2.2 Meanings of "Fraction"

Almost all problems in arithmetic, in one way or another, have to do with *fractions*. Decimals and percents, which will be introduced in Chapters 3 and 4, are other ways of writing fractions. To understand all of these, a basic knowledge of fractions is necessary.

You should know at least three meanings of the word *fraction*.

DEFINITION

**Fraction**

1. *A part of a whole.* This was probably the way you were introduced to fractions. The fraction $\frac{3}{10}$ means we are working with 3 of the 10 equal parts into which a whole has been divided. It is usually read, "3 tenths."
2. *A multiplication.* Just as the expression *3 fives* means 3 times 5, so does $\frac{3}{10}$ (*3 tenths*) mean 3 times 1 tenth, or $3 \times \frac{1}{10}$.
3. *A division.* This is the meaning that you will find most useful in the review of operating with fractions. The fraction $\frac{3}{10}$ represents the *quotient* when 3 is divided by 10, which may also be written: $10\overline{)3}$, $3 \div 10$, and now $\frac{3}{10}$.

A fraction has two terms: the numerator and the denominator. *The numerator, written above the line, tells us how many parts we are considering,* while *the denominator, written below the line, tells us into how many parts the whole has been divided.* For example, in the fraction $\frac{3}{10}$, 3 is the numerator and 10 is the denominator. The 10 tells us that the whole has been divided into 10 equal parts, and the 3 tells us to consider 3 of these parts.

Since you cannot divide by zero, *the denominator can never equal zero.*

It is important to know the following:

DEFINITIONS

A **proper fraction** is a fraction whose *numerator is less than its denominator:* $n < d$. ($<$ means "is less than.")

An **improper fraction** is one where the numerator is greater than the denominator: $n > d$. ($>$ means "is greater than.")

Try the following exercises on fractions. Here is a key to some of the symbols used:

$a > b$ means "$a$ is greater than $b$."

$a < b$ means "$a$ is less than $b$."

$a \neq b$ means "$a$ does *not* equal $b$."

$a \leqslant b$ means "$a$ is less than *or* equal to $b$."

$a \geqslant b$ means "$a$ is greater than *or* equal to $b$."

**Exercises**

True or false?

1. $\frac{(5+2)}{(5+3)} = 7 \times \frac{1}{8}$

2. $\frac{(1+1)}{6} = 2\overline{)6}$ or 3

3. $6 + 1 = 7 \times 1$

4. $\frac{6}{0} = 6$

5. $\frac{(1-1)}{(10-10)} = 0$

6. $\frac{(7+7)}{(7-7)} = 14$

7. $\frac{(a-a)}{b} = 0$ if $b \neq 0$

8. $\frac{(a-a)}{b} = 0$ if $a \neq 0$

9. $\frac{a \times b}{b} = a$ if $b \neq 0$

10. $2 + 2 \neq 2 \times 2$

11. $3 + 2 > 5 \times 1$

12. $\frac{1}{10} < \frac{9}{10}$

13. $7 \leq 5 + 2$

14. $\frac{b}{b} = 1$ if $b \neq 0$

15. $\frac{0}{5} = 0$

16. $\frac{5}{0} =$ no number

17. $\frac{0}{0} =$ no number

18. $\frac{(3+5)}{(3 \times 3)}$ means that we have divided a whole into 9 pieces and that we are talking about 8 of them.

**Answers**

1. True
2. False, for the fraction represents $6\overline{)2}$ or $\frac{2}{6}$, which $= \frac{1}{3}$.
3. True
4. False, for any number *over zero* = *no number*.

**c.** False, for the fraction equals $\frac{0}{0}$ or *no number*.

**6.** False, for $\frac{14}{0}$ equals *no number*.

**7.** True, for zero divided by any non-zero number will equal zero.

**8.** False, for $\frac{(a-a)}{b}$ equals $\frac{0}{b}$ no matter what the value of $a$ is; the value of $\frac{0}{b}$ is 0 if and only if $b \neq 0$.

**9.** True, for $\frac{b}{b}$ will equal 1, if $b \neq 0$, and any number $a$ times 1 will equal that number $a$ as stated in the multiplicative property of *one*.

**10.** False, for $2 + 2 = 4$ and $2 \times 2 = 4$, and therefore, $2 + 2$ does $= 2 \times 2$.

**11.** False, for $3 + 2 = 5$ and $5 \times 1 = 5$, so $3 + 2$ is *not greater than* but *is equal to* $5 \times 1$.

**12.** True, for *1 tenth* is less than *9 tenths*.

**13.** True, for the statement says that 7 is less than *or* equal to $5 + 2$, and 7 *equals* $5 + 2$.

**14.** True, for our properties of *one* tell us that any number divided by itself equals *one*, provided that number $\neq 0$.

**15.** True, for zero divided by any non-zero number equals zero.

**16.** True, for any number divided by zero equals no number.

**17.** True, for even zero cannot be divided by zero.

**18.** True

# 2.3 Multiplication of Fractions and Lowest Terms

To *multiply* two fractions, simply *multiply the two numerators and the two denominators*.

*Example:* $\frac{2}{3} \times \frac{5}{7} = \frac{10}{21}$, because $2 \times 5 = 10$ and $3 \times 7 = 21$.

A little later you will find some shortcuts for this. First, examine a special case of multiplication of fractions. Remember that if any number is multiplied by *one*, its value is not changed. This is also true when *one* appears as a fraction, such as $\frac{5}{5}$.

*Example:* $\frac{2}{3} \times \frac{5}{5} = \frac{10}{15}$

$\frac{10}{15}$ must be the same as $\frac{2}{3}$, since $\frac{2}{3}$ was multiplied by $\frac{5}{5}$, which is 1. Remember that any number times 1 stays unchanged. Since $\frac{10}{15}$ and $\frac{2}{3}$ are the same number, we say that $\frac{2}{3}$ is $\frac{10}{15}$ written in its "lowest terms."

A fraction in lowest terms is one in which the numerator and denominator have no common factor.

> To reduce a fraction to *lowest terms,* divide the numerator and denominator by the same number. When there are no whole numbers except 1 which will divide the numerator and denominator, the fraction is in lowest terms.

*Example:* Reduce $\frac{24}{36}$ to lowest terms.

STEP 1 Find a number which is a factor of both 24 and 36. 4 is a divisor of both numbers.

STEP 2 Divide 24 and 36 each by 4.
$24 \div 4 = 6$
$36 \div 4 = 9$
So $\frac{24}{36} = \frac{6}{9}$.

STEP 3 Is there a number which will divide both 6 and 9? Yes, 3 will do the trick. Divide 6 and 9 each by 3
$6 \div 3 = 2$
and $9 \div 3 = 3$,
so $\frac{6}{9} = \frac{2}{3}$.

$\frac{2}{3}$ is the fraction $\frac{24}{36}$ in its lowest terms.

**Exercises** Reduce each fraction to its lowest terms.

**1.** $\frac{6}{10}$ **2.** $\frac{18}{12}$ **3.** $\frac{36}{9}$ **4.** $\frac{9}{16}$

**Answers**

**1.** $\frac{3}{5}$ **2.** $\frac{3}{2}$ **3.** 4 **4.** $\frac{9}{16}$

Notice that $\frac{9}{16}$ was already in lowest terms.

Before leaving the topic of reducing to lowest terms we should look at two related concepts. A ratio is another name for a fraction, usually used to compare two quantities. For example, the ratio of 6 inches to a foot is $\frac{6}{12}$ (since there are 12 inches in a foot).

The statement that two ratios are equal is called a *proportion*. For example, $\frac{6}{12} = \frac{1}{2}$ is a proportion. Notice that there are four numbers in a proportion. Often three of these will be known with the fourth to be determined.

For instance, let us find the fourth number to make the following a true proportion:

$$\frac{3}{5} = \frac{12}{?}$$

Since 3 must be multiplied by 4 to get 12, the unknown can be found by multiplying 5 by 4:

$$\frac{3}{5} = \frac{12}{20}$$

# 2.4 Mixed Numbers

A *mixed number* is a number made up of a whole number and a fraction. $3\frac{1}{2}$ is a mixed number. It is read "3 and $\frac{1}{2}$," so it must mean $3 + \frac{1}{2}$.

> To change a fraction to a mixed number, the fraction itself must be *improper*. (Remember that the numerator of an improper fraction is larger than the denominator.) Look at the fraction $\frac{16}{3}$, which means $16 \div 3$. If 16 is divided by 3, the quotient is 5 and the remainder is 1. So $\frac{16}{3}$ must be the same as $5\frac{1}{3}$.

*Example:* Change $\frac{37}{5}$ to a mixed number.

$37 \div 5 = 7 + 2$ remainder,

so $\frac{37}{5} = 7\frac{2}{5}$.

**Exercises** Convert these fractions to mixed numbers.

1. $\frac{11}{4}$
2. $\frac{26}{7}$
3. $\frac{91}{13}$
4. $\frac{5}{6}$

**Answers**

1. $2\frac{3}{4}$  2. $3\frac{5}{7}$  3. and

4. are not really mixed numbers, because $\frac{91}{13} = 7$ and $\frac{5}{9}$ has a denominator larger than its numerator.

*Changing a mixed number to a fraction* is easier than changing a fraction to a mixed number. For example, if we wish to convert $2\frac{1}{3}$ to a fraction, we first write $2\frac{1}{3}$ as $2 + \frac{1}{3}$. Remember that 2 is the same as $\frac{2}{1}$ or $\frac{6}{3}$ (multiplying by $\frac{3}{3}$). $\frac{6}{3} + \frac{1}{3} = \frac{7}{3}$. You can check by converting $\frac{7}{3}$ back to a mixed number.

> To change a mixed number to a fraction, multiply the "whole number" part of the mixed number by the "fraction" part. Add this result to the numerator and keep the same denominator.

*Example:* Change $2\frac{1}{3}$ to an improper fraction.

$2 \times 3 = 6$ and $6 + 1 = 7$

So the fraction equal to $2\frac{1}{3}$ is $\frac{7}{3}$.

**Exercises** Change each mixed number to an improper fraction.

1. $7\frac{1}{5}$  2. $3\frac{1}{4}$  3. $12\frac{2}{3}$  4. $4\frac{1}{10}$

**Answers**

1. $\frac{36}{5}$  2. $\frac{13}{4}$  3. $\frac{38}{3}$  4. $\frac{41}{10}$

# 2.5 Multiplication Short Cuts — Cancellation

If we were to multiply $\frac{16}{27} \times \frac{3}{4}$, we would get $\frac{48}{108}$, which could be reduced, with a little hard work, to $\frac{4}{9}$.

However, an easier way to handle the problem would be to observe that the numerator of $\frac{3}{4}$ and the denominator of $\frac{16}{27}$ can both be divided by 3. You can also see that the numerator of $\frac{16}{27}$ and the denominator of $\frac{3}{4}$ can both be divided by 4. Now follow these steps:

STEP 1 Divide 4 into 16 and into 4 as follows:

$$\frac{\overset{4}{\cancel{16}}}{27} \times \frac{3}{\underset{1}{\cancel{4}}}$$

STEP 2 Divide 3 into 27 and into 3 as follows:

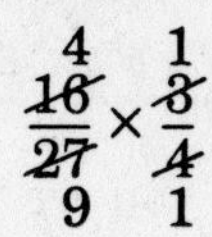

STEP 3 Now multiply numerators and denominators to obtain $\frac{4}{9}$.

This shortcut is called *cancellation*. You are really dividing the same number into both numerator and denominator at the same time, so that the product of the two fractions is in *lowest terms*.

$$\frac{16}{27} \times \frac{3}{4} = \frac{48}{108}$$

THERE MUST BE A BETTER WAY!

$$\frac{\overset{4}{\cancel{16}}}{27} \times \frac{3}{\underset{1}{\cancel{4}}} = \frac{4}{\underset{9}{\cancel{27}}} \times \frac{\overset{1}{\cancel{3}}}{1} = \frac{4}{9}$$

WE HAVE A COMMON FACTOR!

...AND WE HAVE A COMMON FACTOR!

AHH.. THAT WAS SO EASY!

**Exercises** Multiply each of the following pairs of numbers.

(HINT: Convert mixed numbers to improper fractions before multiplying.)

**1.** $\frac{7}{8} \times \frac{5}{6}$

**2.** $14 \times \frac{5}{2}$

**3.** $\frac{25}{24} \times \frac{2}{5}$

**4.** $\frac{8}{9} \times \frac{3}{2}$

**5.** $\frac{10}{11} \times \frac{33}{25}$

**6.** $2\frac{4}{7} \times \frac{1}{6}$

**7.** $7\frac{1}{2} \times 2\frac{2}{3}$

**8.** $16\frac{1}{3} \times 2\frac{1}{7}$

**Answers**

1. $\frac{35}{48}$ 3. $\frac{5}{12}$ 5. $\frac{9}{5}$ 7. 20
2. 35 4. $\frac{4}{3}$ 6. $\frac{3}{7}$ 8. 35

# 2.6 Division of Fractions

If you were asked, "How many quarters are there in 2?" you would say that there were 8. This is really the same problem as $2 \div \frac{1}{4}$, since this asks how many one-fourths there are in 2.

One way to divide one fraction by another is illustrated in the following example:

$$\frac{2}{9} \div \frac{2}{3} = \frac{2 \div 2}{9 \div 3} = \frac{1}{3}$$

This method, dividing numerator by numerator and denominator by denominator, will work nicely only when each is a multiple of the other. The method shown above cannot be applied directly to $2\frac{1}{2} \div \frac{3}{4}$, which is $\frac{5}{2} \div \frac{3}{4}$. However, since $\frac{5}{2} = \frac{10}{4}$, we write $\frac{10}{4} \div \frac{3}{4}$, which is $\frac{(10 \div 3)}{1}$. Thus the answer is $\frac{10}{3}$ or $3\frac{1}{3}$.

An easier way to divide fractions is to use the idea of a *reciprocal*. What number multiplied by $\frac{3}{4}$ will give us a product of 1? $\frac{3}{4} \times ? = 1$. The number is $\frac{4}{3}$, since $\frac{3}{4} \times \frac{4}{3} = 1$. We say that if two fractions have a product which is 1, the fractions are *reciprocals* of each other. Find reciprocals for $\frac{2}{3}$, $\frac{7}{8}$, $4\frac{1}{2}$, and 7. Answers: $\frac{3}{2}$, $\frac{8}{7}$, $\frac{2}{9}$ (since $4\frac{1}{2} = \frac{9}{2}$) and $\frac{1}{7}$ (since $7 = \frac{7}{1}$). An often used name for a reciprocal of a number is the multiplicative inverse of the number.

If we apply this concept of a reciprocal to the division of fractions, a simple method of division will be discovered.

*Example:* $4\frac{1}{2} \div 1\frac{3}{4}$

STEP 1 Rewrite the problem using improper fractions.

$$\frac{9}{2} \div \frac{7}{4}$$

STEP 2 Rewrite this result as a *complex fraction.* (A complex fraction has a fractional numerator or denominator — or both.)

$$\frac{\frac{9}{2}}{\frac{7}{4}}$$

STEP 3 Find the reciprocal of the denominator.

$$\frac{4}{7}$$

STEP 4 Recall that any number divided by itself is 1, and that multiplying a number by 1 does not change it.

$$\frac{4}{7} \div \frac{4}{7} \text{ or } \frac{\frac{4}{7}}{\frac{4}{7}} = 1$$

STEP 5 Multiply the complex fraction $\frac{\frac{9}{2}}{\frac{7}{4}}$ by $\frac{\frac{4}{7}}{\frac{4}{7}}$.

$$\frac{\frac{9}{2}}{\frac{7}{4}} \times \frac{\frac{4}{7}}{\frac{4}{7}} = \frac{\frac{9}{2} \times \frac{4}{7}}{1} = \frac{\frac{18}{7}}{1} \text{ or } \frac{18}{7}, \text{ which is } 2\frac{4}{7}.$$

$\frac{7}{4}$ was the *divisor* in the original problem, but now you can see that $\frac{9}{2} \div \frac{7}{4}$ is the same as $\frac{9}{2} \times \frac{4}{7}$, which is $2\frac{4}{7}$.

To *divide* one fraction by another, simply *multiply the first fraction by the reciprocal of the second.*

$$\frac{a}{b} \div \frac{c}{d} \text{ is the same as } \frac{a}{b} \times \frac{d}{c} \text{ or } \frac{a \times d}{b \times c}.$$

*Examples:* 1. $\frac{27}{4} \div \frac{3}{2}$

Invert the divisor, "cancel," and multiply.

$$\frac{\overset{9}{\cancel{27}}}{\underset{2}{\cancel{4}}} \times \frac{\overset{1}{\cancel{2}}}{\underset{1}{\cancel{3}}} = \frac{9}{2} \text{ or } 4\frac{1}{2}$$

2. $4\frac{1}{5} \div 2\frac{1}{3}$

First convert each mixed number to an improper fraction.

$$\frac{21}{5} \div \frac{7}{3}$$

Now invert the divisor, "cancel," and multiply.

$$\frac{\overset{3}{\cancel{21}}}{5} \times \frac{3}{\underset{1}{\cancel{7}}} = \frac{9}{5} \text{ or } 1\frac{4}{5}$$

**Exercises** Perform the indicated divisions.

1. $\frac{3}{5} \div \frac{3}{4}$

2. $2\frac{2}{3} \div 1\frac{2}{3}$

3. $4\frac{1}{2} \div \frac{3}{4}$

**4.** $6 \div 1\frac{1}{2}$ **5.** $12\frac{1}{2} \div 2\frac{1}{2}$ **6.** $\frac{3}{5} \div 3$

Answers

**1.** $\frac{4}{5}$ **2.** $\frac{8}{5} = 1\frac{3}{5}$ **3.** 6 **4.** 4 **5.** 5 **6.** $\frac{1}{5}$

# 2.7 Addition of Fractions

Since fractions are parts of a whole, and the denominator indicates the number of parts, we can add only fractions which have the same denominator.

Consider $\frac{3}{7} + \frac{2}{7}$. If we read this aloud, it says, "3 sevenths plus 2 sevenths." Just as if we said "3 dollars plus 2 dollars is 5 dollars," we say "3 sevenths plus 2 sevenths is 5 sevenths."

Thus, $\frac{3}{4} + \frac{1}{4} = \frac{4}{4}$ or 1, $\frac{1}{5} + \frac{3}{5} = \frac{4}{5}$, and $\frac{1}{9} + \frac{5}{9} = \frac{6}{9}$ or $\frac{2}{3}$.

> To add fractions with the same denominator, *add their numerators* and *preserve the denominator.*

If the fractions do not have the same denominators, find a way to change them so that the denominators are the same.

*Examples:* **1.** $\frac{1}{2} + \frac{1}{4}$

First, find an equivalent fraction for $\frac{1}{2}$ which has the same denominator as $\frac{1}{4}$. This can be done by multiplying $\frac{1}{2}$ by $\frac{2}{2}$ — that is, by 1 — to obtain $\frac{2}{4}$. Then

$$\frac{1}{2} + \frac{1}{4} \text{ becomes } \frac{2}{4} + \frac{1}{4} = \frac{3}{4}$$

**2.** $\frac{1}{2} + \frac{1}{3}$

Now there is no convenient way to change one fraction so its denominator will be the same as the other. Both denominators must be changed. One way to decide what to multiply by to make sure each fraction has the same denominator is to multiply the two denominators together. In this case $2 \times 3 = 6$. Now we know that $\frac{1}{2}$ must be multiplied by $\frac{3}{3}$ and $\frac{1}{3}$ by $\frac{2}{2}$. Then the denominator in each case is 6.

$$\frac{1}{2} + \frac{1}{3} = \left(\frac{1}{2} \times \frac{3}{3}\right) + \left(\frac{1}{3} \times \frac{2}{2}\right)$$
$$= \frac{3}{6} + \frac{2}{6} = \frac{5}{6}$$

> To add two (or more) mixed numbers, *add the whole number parts first;* then *add the fractions.*

*Examples:* **1.** $5\frac{1}{4} + 3\frac{2}{3}$

Rewrite $(5 + 3) + (\frac{1}{4} + \frac{2}{3})$

Add $5 + 3 = 8$

and $\frac{1}{4} + \frac{2}{3} = \left(\frac{1}{4} \times \frac{3}{3}\right) + \left(\frac{2}{3} \times \frac{4}{4}\right)$

$= \frac{3}{12} + \frac{8}{12}$

$= \frac{11}{12}$

Thus $5\frac{1}{4} + 3\frac{2}{3} = 8\frac{11}{12}$

**2.** $3\frac{2}{3} + 7\frac{3}{5} + 4\frac{1}{4}$

Rewrite $(3 + 7 + 4) + (\frac{2}{3} + \frac{3}{5} + \frac{1}{4})$

Add $3 + 7 + 4 = 14$

$\frac{2}{3} + \frac{3}{5} + \frac{1}{4}$ must be combined:

$\left(\frac{2}{3} \times \frac{20}{20}\right) + \left(\frac{3}{5} \times \frac{12}{12}\right) + \left(\frac{1}{4} \times \frac{15}{15}\right)$

$= \frac{40}{60} + \frac{36}{60} + \frac{15}{60}$

$= \frac{91}{60}$ or $1\frac{31}{60}$

Thus $3\frac{2}{3} + 7\frac{3}{5} + 4\frac{1}{4} = 14 + 1\frac{31}{60} = 15\frac{31}{60}$

LOWEST COMMON DENOMINATOR

REMEMBER US??

In this example, the denominator 60 is called the <u>common denominator</u>, for we changed each fraction to have 60 as a denominator. Again, an easy way to find a common denominator is to multiply the three denominators together. Here, $3 \times 5 \times 4 = 60$.

*For simplicity it is desirable to find the smallest or <u>lowest</u> common denominator.*

To illustrate the concept of *lowest common denominator,* consider the following example:

$$\frac{3}{4} + \frac{1}{2} = ?$$

There are many *common* denominators — 4, 8, 16, 32, 64 and so on — because 4 and 2 are divisors of each of these numbers. The *lowest* common denominator, however, is 4. Rewrite $\frac{1}{2}$ as $\frac{2}{4}$ and add:

$$\frac{3}{4} + \frac{2}{4} = \frac{5}{4} \text{ or } 1\frac{1}{4}$$

**Exercises** Find the sum in each case.

1. $\frac{3}{4} + \frac{1}{8}$
2. $\frac{2}{3} + \frac{1}{6}$
3. $\frac{2}{3} + \frac{1}{5}$
4. $\frac{5}{6} + \frac{3}{4}$
5. $2\frac{3}{4} + 3\frac{3}{8}$
6. $3\frac{2}{3} + 6\frac{5}{6}$

**Answers** **1.** $\frac{7}{8}$ **2.** $\frac{5}{6}$ **3.** $\frac{13}{15}$ **4.** $1\frac{7}{12}$ **5.** $6\frac{1}{8}$ **6.** $10\frac{1}{2}$

# 2.8 Subtraction of Fractions

Subtraction of fractions follows the same pattern as addition. If the denominators are the *same*, keep the common denominator and *subtract the second numerator from the first.* If the denominators are *different, change the fractions so that they have the same denominator.*

*Examples:* **1.** $\frac{3}{4} - \frac{1}{4} = \frac{2}{4} = \frac{1}{2}$

2. $\frac{7}{8} - \frac{1}{4} = \frac{7}{8} - \frac{2}{8} = \frac{5}{8}$

3. $\frac{2}{3} - \frac{1}{4} = \left(\frac{2}{3} \times \frac{4}{4}\right) - \left(\frac{1}{4} \times \frac{3}{3}\right) = \frac{8}{12} - \frac{3}{12} = \frac{5}{12}$

In subtracting mixed numbers, it may sometimes be necessary to "borrow" as in the subtraction of whole numbers.

*Examples:* 1. $5 - \frac{1}{3}$

5 is the same as $4 + 1$ or $4 + \frac{3}{3}$.

The problem becomes $4 + \frac{3}{3} - \frac{1}{3}$ or $4\frac{2}{3}$.

2. $7\frac{1}{2} - 4\frac{1}{3}$

Rewrite
$$\begin{array}{r} 7\frac{1}{2} \\ -\ 4\frac{1}{3} \\ \hline \end{array}$$

Subtract $7 - 4 = 3$

and $\frac{1}{2} - \frac{1}{3} = \frac{3}{6} - \frac{2}{6} = \frac{1}{6}$

Thus $7\frac{1}{2} - 4\frac{1}{3} = 3\frac{1}{6}$

3. $11\frac{1}{5} - 7\frac{2}{3}$

Rewrite
$$\begin{array}{r} 11\frac{1}{5} \\ -\ 7\frac{2}{3} \\ \hline \end{array}$$

The fraction $\frac{2}{3}$ is larger than the fraction $\frac{1}{5}$. Therefore, "borrow" 1 from 11 and add it to $\frac{1}{5}$ in the form of $\frac{5}{5}$.

$1 + \frac{1}{5} = \frac{5}{5} + \frac{1}{5} = \frac{6}{5}$

Rewrite
$$\begin{array}{r} 10\frac{6}{5} \\ -\ 7\frac{2}{3} \\ \hline \end{array}$$

Subtract $10 - 7 = 3$

and $\frac{6}{5} - \frac{2}{3} = \left(\frac{6}{5} \times \frac{3}{3}\right) - \left(\frac{2}{3} \times \frac{5}{5}\right)$

$= \frac{18}{15} - \frac{10}{15} = \frac{8}{15}$

Thus $11\frac{1}{5} - 7\frac{2}{3} = 3\frac{8}{15}$.

**Exercises**

1. $\frac{5}{8} - \frac{1}{8}$
2. $\frac{3}{4} - \frac{3}{4}$
3. $\frac{5}{8} - \frac{1}{4}$
4. $\frac{7}{16} - \frac{3}{8}$
5. $\frac{2}{3} - \frac{2}{5}$
6. $4\frac{1}{4} - 3\frac{15}{16}$
7. $14\frac{1}{8} - 13\frac{3}{4}$
8. $11\frac{1}{16} - 4\frac{3}{32}$

**Answers**

1. $\frac{1}{2}$
2. 0
3. $\frac{3}{8}$
4. $\frac{1}{16}$
5. $\frac{4}{15}$
6. $\frac{5}{16}$
7. $\frac{3}{8}$
8. $6\frac{31}{32}$

# Trial Test — Fractions

**1.** The reciprocal of $\frac{11}{4}$ is

(1) $\frac{11}{4}$
(2) $\frac{4}{11}$
(3) $2\frac{3}{4}$
(4) 1
(5) $\frac{1}{4}$

**1.** 1 2 3 4 5
|| || || || ||

**2.** A fraction whose value is the same as $\frac{3}{4}$ is

(1) $\frac{3}{2}$
(2) $\frac{4}{4}$
(3) $\frac{15}{20}$
(4) $\frac{9}{15}$
(5) $\frac{6}{9}$

**2.** 1 2 3 4 5
|| || || || ||

**3.** Change $5\frac{1}{4}$ to an improper fraction.

(1) $\frac{51}{4}$
(2) $\frac{20}{4}$
(3) $\frac{21}{2}$
(4) 21
(5) $\frac{21}{4}$

**3.** 1 2 3 4 5
|| || || || ||

**4.** Change $\frac{31}{4}$ to a mixed number.

(1) $7\frac{3}{4}$
(2) 7
(3) $\frac{3}{4}$
(4) $7\frac{1}{4}$
(5) 8

**4.** 1 2 3 4 5

**5.** $\frac{5}{7} + \frac{1}{7} =$

(1) $\frac{6}{14}$
(2) $\frac{6}{7}$
(3) $\frac{5}{49}$
(4) $\frac{5}{14}$
(5) $\frac{5}{7}$

**5.** 1 2 3 4 5

**6.** The fraction $\frac{28}{84}$, when written in lowest terms, is

(1) $\frac{14}{42}$
(2) $\frac{4}{12}$
(3) $\frac{7}{21}$
(4) 3
(5) $\frac{1}{3}$

**6.** 1 2 3 4 5

**7.** Find the product of $\frac{5}{6}$ and $\frac{3}{7}$.

(1) $\frac{8}{13}$
(2) $\frac{15}{21}$
(3) $\frac{15}{14}$
(4) $\frac{5}{14}$
(5) $\frac{15}{13}$

**7.** 1 2 3 4 5

**8.** 16 is what part of 96?

(1) $\frac{1}{6}$
(2) $\frac{1}{8}$
(3) $\frac{1}{16}$
(4) $\frac{16}{24}$
(5) $\frac{1}{12}$

**8.** 1 2 3 4 5

**9.** Add $3\frac{1}{4} + 5\frac{1}{8}$.

(1) $\frac{3}{8}$
(2) 8
(3) $8\frac{3}{32}$
(4) $8\frac{3}{8}$
(5) $8\frac{1}{32}$

**9.** 1 2 3 4 5

**10.** Find the difference between $\frac{3}{4}$ and $\frac{1}{4}$.

(1) $\frac{2}{8}$
(2) $\frac{2}{0}$
(3) $\frac{1}{2}$
(4) $\frac{3}{16}$
(5) 1

10. 1 2 3 4 5

**11.** $7\frac{1}{2} + 5\frac{1}{3} + 3\frac{1}{4} =$

(1) 15
(2) $15\frac{1}{9}$
(3) $15\frac{1}{25}$
(4) $16\frac{1}{12}$
(5) $15\frac{3}{9}$

11. 1 2 3 4 5

**12.** Find the quotient: $\frac{7}{8} \div \frac{3}{4}$

(1) 2
(2) $\frac{7}{6}$
(3) $\frac{21}{32}$
(4) $\frac{6}{7}$
(5) $\frac{32}{21}$

12. 1 2 3 4 5

**13.** $17 - \frac{15}{16} =$

(1) $\frac{1}{16}$
(2) $17\frac{1}{16}$
(3) $16\frac{1}{16}$
(4) $\frac{2}{16}$
(5) 15

13. 1 2 3 4 5

**14.** $\frac{7}{10} \times \frac{15}{16} \times \frac{20}{21} =$

(1) $\frac{51}{37}$
(2) $\frac{5}{8}$
(3) $\frac{8}{5}$
(4) $\frac{22}{47}$
(5) $\frac{100}{210}$

14. 1 2 3 4 5

**15.** How many yards of material are needed to make 5 dresses if each requires $2\frac{3}{4}$ yards?

(1) 15 yards
(2) 10 yards
(3) 14 yards
(4) $13\frac{3}{4}$ yards
(5) 23 yards

15. 1 2 3 4 5

## SOLUTIONS

1. (2) The reciprocal is the number which when multiplied by $\frac{11}{4}$ yields 1.

$$\frac{11}{4} \times \frac{4}{11} = 1$$

2. (3) $\frac{3}{4} \times \frac{5}{5} = \frac{15}{20}$. Since we multiply by $\frac{5}{5}$, which is the same as 1, the number $\frac{3}{4}$ is left unchanged.

3. (5) $5\frac{1}{4}$ is the same as $\frac{5}{1} + \frac{1}{4} = \frac{20}{4} + \frac{1}{4}$ or $\frac{21}{4}$.

4. (1) $\frac{31}{4}$ is the same as $31 \div 4$.

$$4\overline{)31}\ \ 7\text{ R }3 \qquad \text{so } \frac{31}{4} = 7\frac{3}{4}.$$

5. (2) $\frac{5}{7} + \frac{1}{7} = \frac{6}{7}$. Keep the common denominator and add the numerators.

6. (5) $\frac{28}{84}$ stays the same when the numerator and denominator are divided by the same number.

$$\frac{28}{84} \div \frac{28}{28} = \frac{1}{3}$$

7. (4) $\frac{5}{6} \times \frac{3}{7} = \frac{15}{42} = \frac{5}{14}$. Remember that the word *product* means the answer in multiplication.

8. (1) To find what part 16 is of 96, compare them by division. $\frac{16}{96}$ can be reduced to $\frac{1}{6}$.

9. (4) $3\frac{1}{4} + 5\frac{1}{8} = (3 + 5) + (\frac{1}{4} + \frac{1}{8})$

$$= 8 + (\frac{2}{8} + \frac{1}{8})$$

$$= 8\frac{3}{8}$$

10. (3) The word *difference* means subtract.

$$\frac{3}{4} - \frac{1}{4} = \frac{2}{4}, \text{ or } \frac{1}{2}$$

11. (4) $7\frac{1}{2} + 5\frac{1}{3} + 3\frac{1}{4} = (7 + 5 + 3) + (\frac{1}{2} + \frac{1}{3} + \frac{1}{4})$

$$= 15 + (\frac{1}{2} + \frac{1}{3} + \frac{1}{4})$$

The lowest common denominator is 12.

$$= 15 + (\frac{6}{12} + \frac{4}{12} + \frac{3}{12})$$

$$= 15 + \frac{13}{12}, \text{ or } 15 + 1\frac{1}{12}$$

$$= 16\frac{1}{12}$$

12. (2) $\frac{7}{8} \div \frac{3}{4}$ is the same as $\frac{7}{8}$ multiplied by the reciprocal of $\frac{3}{4}$, which is $\frac{4}{3}$.

$$\frac{7}{8} \times \frac{4}{3} = \frac{7}{6}$$

13. (3) $17 - \frac{15}{16}$ must be rewritten borrowing 1 from 17. 1 may be written as $\frac{16}{16}$. Thus $17 - \frac{15}{16}$ becomes $16\frac{16}{16} - \frac{15}{16} = 16\frac{1}{16}$.

14. (2) $\frac{7}{10} \times \frac{15}{16} \times \frac{20}{21} = \frac{5}{8}$ with repeated use of the shortcut, cancellation.

15. (4) If 5 dresses are required and $2\frac{3}{4}$ yards are needed for each, the operation must be multiplication. Thus $5 \times 2\frac{3}{4}$ becomes

$$\frac{5}{1} \times \frac{11}{4} = \frac{55}{4}, \text{ or } 13\frac{3}{4} \text{ yards.}$$

## EVALUATION CHART

Multiply the number of correct answers by 100 and divide by the number of questions in the Trial Test to arrive at your score. ____________

| Score | Rating |
|---|---|
| 90–100 | Excellent |
| 80–89 | Very Good |
| 70–79 | Average |
| 60–69 | Passing |
| 59 or Below | Very Poor |

# Chapter 3

# DECIMAL FRACTIONS

# 3.1 Meaning of "Decimal Fraction"

Now that we have looked at *common fractions,* let's turn our attention to decimal fractions. One situation where we all see and use decimal fractions involves money. What does $1.98 really mean? You know it means 1 *dollar* and 98 *cents*. But cents, after all, are fractional parts of a dollar. Decimal fractions are just a different way of writing fractions which have special denominators.

> DEFINITION
>
> **Decimals** are special fractions with denominators which are powers of ten.

Powers of ten are easy to remember:

$$10^1 = 10,\ 10^2 = 10 \times 10 = 100,\ 10^3 = 10 \times 10 \times 10 = 1{,}000, \text{ etc.}$$

The exponent tells you how many zeros there are in the power of ten. For example, $10^6 = 1{,}000{,}000$ (one million).

Every decimal fraction has three parts: a whole number (often zero), followed by a decimal point and one or more whole numbers. For example, a simple decimal fraction is 0.7, which is the same as a common fraction with numerator 7 and denominator 10, so 0.7 is the same as $\frac{7}{10}$. The decimal fraction 0.07 still has a numerator of 7, but the denominator is $10^2$ or 100, so 0.07 is the same as $\frac{7}{100}$. Notice that the number of places to the right of the decimal point is the same as the number of *zeros* in the denominator of the fraction equivalent ($0.\underline{07}, \frac{7}{1\underline{00}}$) and also the same as the power of ten in the denominator ($0.\underline{07}, \frac{7}{10^{\underline{2}}}$).

*Examples:*

$0.7 = \frac{7}{10^1}$ or $\frac{7}{10}$ (seven tenths)

$0.07 = \frac{7}{10^2}$ or $\frac{7}{100}$ (seven hundredths)

$0.007 = \frac{7}{10^3}$ or $\frac{7}{1{,}000}$ (seven thousandths)

$0.0007 = \frac{7}{10^4}$ or $\frac{7}{10{,}000}$ (seven ten thousandths)

Think about 0.7 again. This means $\frac{7}{10}$. Now look at 0.70. This means a numerator of 70 and denominator of $10^2$ or 100. Thus 0.70 means $\frac{70}{100}$, which is the same as $\frac{7}{10}$ or 0.7. In the same way, 0.700 means $\frac{700}{1{,}000}$ or $\frac{7}{10}$. It's easy to see that adding zeros to the numbers to the right of the decimal point doesn't change the value of the decimal fraction.

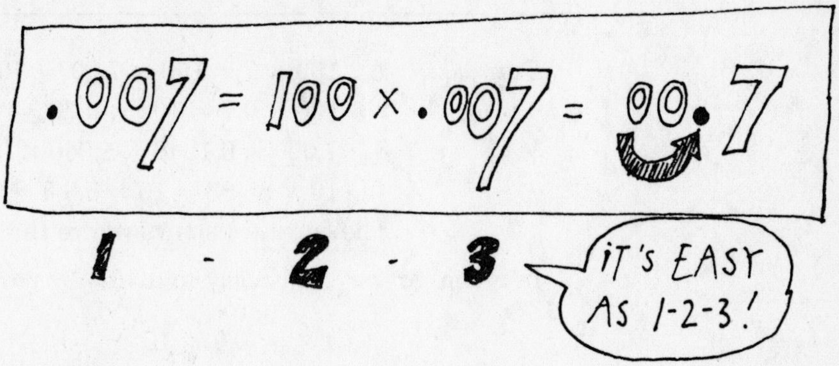

CAUTION: Adding zeros to the *left* of the decimal point will change the value of the fraction: 7. ≠ 70, but 0.7 does = 0.70.

Read some more difficult decimal fractions just to understand the principle.

0.23 means $\frac{23}{100}$

0.205 means $\frac{205}{1{,}000}$

1.3 means $1\frac{3}{10}$ or $\frac{13}{10}$

21.43 means $21\frac{43}{100}$ or $\frac{2{,}143}{100}$

# 3.2 Multiplying and Dividing Decimals by Powers of Ten

Before going on, it's a good idea to review the idea of *place* or *place value*. In the number 21.43, the 2 is in the *ten's place*, the 1 is in the *unit's place*, the 4 is in the *tenth's place* and the 3 is in the *hundredth's place*.

21.43

tens — units — tenths — hundredths

Think about 100 × 0.7, which means $\frac{100}{1} \times \frac{7}{10}$, or 70. Notice that there are *2* zeros in 100 and that the decimal point in the solution has been moved *2* places to the right. This suggests an easy rule:

To *multiply* a decimal by a power of ten, simply move the decimal point to the *right* the same number of places as there are zeros in the power of ten.

*Examples:*
**1.** $1\underline{00} \times 7 = 100 \times 7.\underline{00}_{\wedge} = 700$
**2.** $1\underline{00} \times 0.7 = 100 \times 0.\underline{70}_{\wedge} = 70$
**3.** $1,\underline{000} \times 0.1346 = 1,000 \times 0.\underline{134}_{\wedge}6 = 134.6$
**4.** $1\underline{0} \times \$5.83 = 10 \times 5.\underline{8}_{\wedge}3 = \$58.30$

The carat ( ) shows where the decimal point is being moved.

In a similar way, it is easy to divide by powers of ten.

$$0.7 \div 10 = \frac{7}{10} \times \frac{1}{10} = \frac{7}{100}$$

$\frac{7}{100}$ is written 0.07 as a decimal.

Thus, the decimal point in 0.7 has moved one place to the *left*.

To *divide* a decimal by a power of ten, move the decimal point to the *left* the same number of places as there are zeros in the power of ten.

*Examples:*
**1.** $0.7 \div 1\underline{00} = {}_{\wedge}\underline{00}.7 \div 100 = .007$
**2.** $158.7 \div 1\underline{0} = 15_{\wedge}\underline{8}.7 \div 10 = 15.87$
**3.** $25.6 \div 1,\underline{000} = {}_{\wedge}\underline{025}.6 \div 1,000 = .0256$

# 3.3 Rounding Off Decimals

Many mathematical problems require a rounding off process to reach an answer. Think about 3.7 yards. Is this closer to 3 yards or to 4 yards? $3.7 = 3\frac{7}{10}$, which is closer to 4 yards. How about 3.5 yards? This is exactly midway between 3 and 4 yards. We must make an agreement on rounding off a number of this type. We agree that 3.5 yards will be rounded off to 4 yards.

Now round off 3.346 to the nearest tenth. This lies between 3.3 and 3.4, but is closer to 3.3.

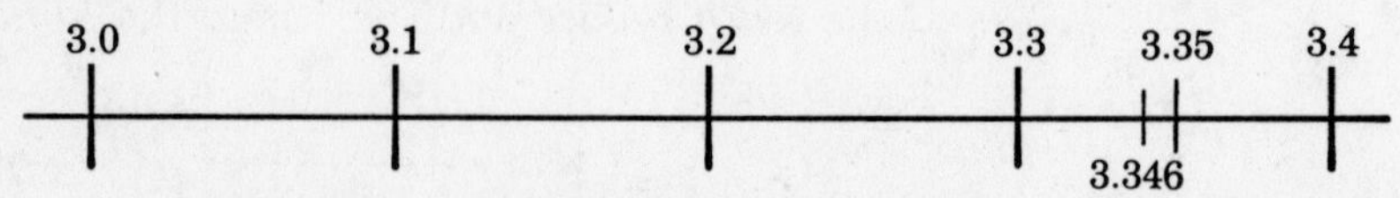


CAUTION: A common mistake in rounding off is to start from the right; in this case the 6 in the thousandth's place will round the 3.346 off to 3.35. This causes our rounded off number to become 3.4 instead of 3.3.

> To round off a decimal to the nearest *tenth*, look at the number in the *hundredth's* place (just to the right of the tenths). If it is 5 or more, round off upward. If it is less than 5, drop it and all numbers following it.

*The same rule applies to rounding off all decimals.* Look at the number to the right of the number in the tenth's place. Is it 5 or more? If it is, round off upward.

*Example:* Round off $37.246 to the nearest *cent* (hundredth).

37.2④6 → hundredth's place

6 is more than 5, so round off upward to $37.25.

**Exercises**

Try the following problems to make sure you understand rounding off.

1. Round off each of these to the nearest tenth:
   **a.** 0.26 **b.** 0.728 **c.** 4.146 **d.** 3.1416
2. Round off each of these to the nearest hundredth:
   **a.** 3.1416 **b.** 8.2763 **c.** 17.0156 **d.** 8.006
3. Round off each of these to the nearest thousandth:
   **a.** 0.8146 **b.** 5.0371 **c.** 3.1416 **d.** 6.0028
4. Round off each of these to the nearest cent (hundredth):
   **a.** $1.432 **b.** $27.486 **c.** $28.997 **d.** $45.466

**Answers**

1. **a.** 0.3 **b.** 0.7 **c.** 4.1 **d.** 3.1
2. **a.** 3.14 **b.** 8.28 **c.** 17.02 **d.** 8.01
3. **a.** 0.815 **b.** 5.037 **c.** 3.142 **d.** 6.003
4. **a.** $1.43 **b.** $27.49 **c.** $29.00 **d.** $45.47

# 3.4 Adding Decimals

Remember that to add two fractions you must have a common denominator. Also, keep in mind that 3.2 is the same as 3.20 or 3.200. Adding zeros after the decimal point does not change the decimal.

> Adding decimals is just as easy as adding whole numbers. Just remember to *keep the decimal points directly under one another.*

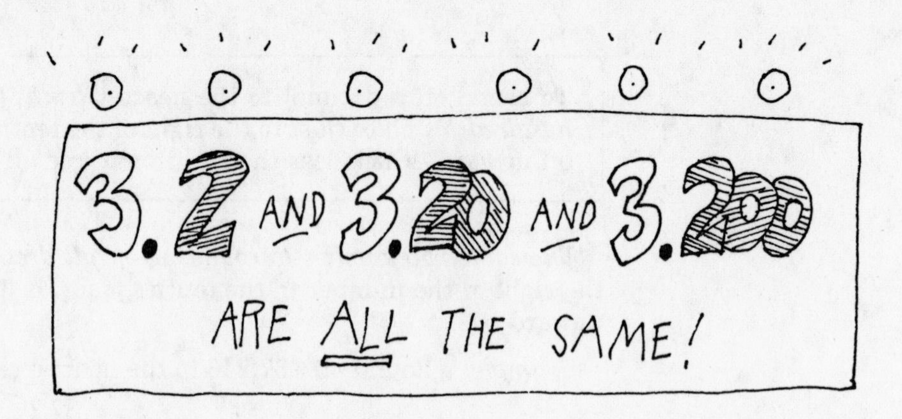


*Examples:* **1.** Add these decimals:

$\$2.25 + \$3.15 + \$12.87$

Just arrange them vertically (up and down) making sure to keep the decimal points under one another.

$$\begin{array}{r} \$2.25 \\ 3.15 \\ +\ 12.87 \\ \hline \$18.27 \end{array}$$

**2.** Add the following decimals:

$3.12 + 14.3 + 205.6 + 0.0324$

First, put them in a column with the decimal points in a line, directly beneath each other. Then add, making sure to put the decimal point in the answer.

$$\begin{array}{r} 3.12 \\ 14.3 \\ 205.6 \\ +\ \ 0.0324 \\ \hline 223.0524 \end{array}$$

If you find it more convenient, add zeros to keep the same number of places, as shown at the right.

$$\begin{array}{r} 3.1200 \\ 14.3000 \\ 205.6000 \\ +\ \ 0.0324 \\ \hline 223.0524 \end{array}$$

**Exercises**

1. 0.03 + 0.14
2. 2.012 + 3.4 + 6.87
3. 25 + 5 + 0.5 + 0.05
4. \$0.71 + \$0.43 + \$2.47 + \$7.36

**Answers** **1.** 0.17 **2.** 12.282 **3.** 30.55 **4.** \$10.97

# 3.5 Subtracting Decimals

The same ideas apply for subtracting decimals as for adding decimals.

> To *subtract* one decimal from another, line up the numbers so that one decimal point is *directly below* the other. Then subtract as you would with whole numbers.

*Examples:* **1.** To subtract \$3.79 from \$5.28 line up the decimal points and perform the subtraction.

$$\begin{array}{r} \$5.28 \\ -\ 3.79 \\ \hline \$1.49 \end{array}$$

**2.** Subtract 3.413 from 8.6. You will find it convenient to add two zeros to 8.6.

$$\begin{array}{r} 8.600 \\ -3.413 \\ \hline 5.187 \end{array}$$

*It's always a good idea to check your answer in subtraction by using addition.* In this case add 3.413 and 5.187 to make sure you get 8.600.

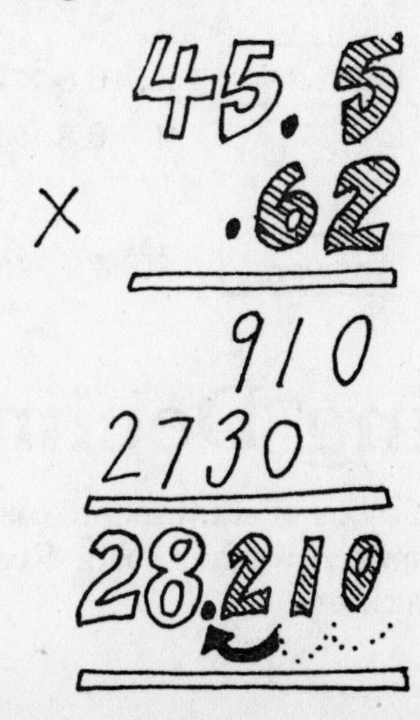

Try the following problems to make sure you understand subtraction of decimals.

**Exercises**

1. $\begin{array}{r} 0.36 \\ \underline{0.027} \end{array}$
2. $\begin{array}{r} 2.4 \\ \underline{0.38} \end{array}$
3. $9 - 0.06$
4. $\$4 - \$1.38$

**Answers**

1. 0.333
2. 2.02
3. 8.94
4. $2.62

# 3.6 Multiplying Decimals

Think about multiplying $0.7 \times 0.3$. This is the same as $\frac{7}{10} \times \frac{3}{10}$, or $\frac{21}{100}$, which is 0.21. Although 0.7 and 0.3 each have *one* decimal place, there are two decimal places in 0.21. This suggests an easy rule to follow in multiplying decimals.

> To multiply two decimals, first multiply them as if they were whole numbers. Then add up their decimal places and insert a decimal point *the same number of places* from the *right* in the solution.

*Example:*

$$\begin{array}{rl} 3.24 & \text{2 decimal places} \\ \times\ \ 1.7 & \underline{+1}\ \text{decimal place} \\ \hline 2\,268 & \\ \underline{3\,24\ \ } & \\ 5.508 & \text{3 decimal places.} \end{array}$$

5.508 is the answer. There are *two* decimal places in 3.24 and *one* decimal place in 1.7. So count off *three* decimal places in the answer.

**Exercises**

1. $0.8 \times 32$
2. $0.8 \times 3.2$
3. $0.8 \times 0.32$
4. $0.43 \times 5.21$

**Answers**

1. 25.6
2. 2.56
3. 0.256
4. 2.2403

# 3.7 Dividing Decimals

If a box of six washers costs $0.24, how much is each one? Clearly $24 \div 6 = 4$, so each washer costs 4 cents or $0.04. Look at the question as a division of decimals.

$$6\overline{)\,0.24}\quad = 0.04$$

Notice that the decimal point in the answer is placed *directly above* the decimal point in the number being divided, or dividend.

*Examples:*

1. $$\begin{array}{r} 0.007 \\ 7\overline{)0.049} \end{array}$$

2. $$\begin{array}{r} 0.12 \\ 12\overline{)1.44} \\ \underline{1\,2} \\ 24 \\ \underline{24} \end{array}$$

3. $$\begin{array}{r} 0.301 \\ 17\overline{)5.117} \\ \underline{5\,1}\phantom{17} \\ 17 \\ \underline{17} \end{array}$$

> To divide a *decimal* by a *whole number,* all you have to do is keep the decimal point in the solution *directly over* the decimal point in the dividend (the number being divided).

What about dividing one decimal by another? Think about 1.8 ÷ 0.2. This could be written as $\frac{1.8}{0.2}$. How can we change the problem so that the divider (0.2) is a whole number? We know that multiplying a number by 1 doesn't change it. Thus,

$$\frac{1.8}{0.2} \times \frac{10}{10} = \frac{18}{2} \text{ or } 9.$$

*Example:* 32.3 ÷ 0.17

Multiply each number by 100, so that the divisor is a whole number. You must add a zero.

$$0.17_{\wedge}\overline{)32.30_{\wedge}}$$

$$\begin{array}{r} 190. \\ 17\overline{)3230.} \\ \underline{17}\phantom{30.} \\ 153\phantom{0.} \\ \underline{153}\phantom{0.} \\ 0\phantom{0.} \end{array}$$

> To divide one decimal by another, move the decimal point enough places *to the right* in both numbers so that the *divisor becomes a whole number.*

*Examples:* 1. 0.00108 ÷ 0.012

Move the decimal point three places in each number (you are multiplying by 1,000).

$$\begin{array}{r} 0.09 \\ 0.012_{\wedge}\overline{)0.001_{\wedge}08} \\ \underline{1\,08} \end{array}$$

**2.** 291.25 ÷ 1.25
Move the decimal point two places in each number.

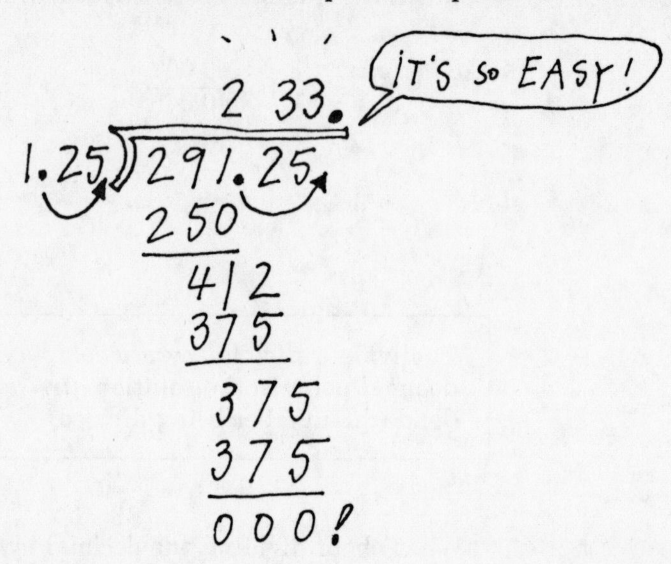


When dividing decimals the chances are that the answer will not come out even as in the examples above. When that happens, you can add zeros *after* the decimal point and carry out the division to any desired accuracy.

*Example:* Find the quotient of 37.2 ÷ 0.07 correct to the nearest tenth.
Add a zero so that the decimal point can be moved two places to the right:

```
0.07‸ ) 37.20‸

       531.42
  7 ) 3720.00
      35
      22
      21
       10
        7
        30
        28
         20
         14
```

531.4 is the answer correct to the nearest tenth.

**Exercises** Find the quotient to the nearest hundredth (if there is a remainder).

**1.** 2.8 ) 0.92
**2.** 0.28 ) 4.8
**3.** 0.08 ) 0.7
**4.** 3.1 ) 1.5
**5.** 0.012 ) 0.3248

**Answers** **1.** 0.33 **2.** 17.14 **3.** 8.75 **4.** 0.48 **5.** 27.07

# 3.8 Converting a Fraction to a Decimal

A fraction is just a different way to indicate division, so the rule is:

> To convert a fraction to a decimal, simply *divide* the numerator by the denominator and carry your answer to any desired degree of accuracy.

*Examples:* Convert each of the fractions to decimals, rounding off to the nearest hundredth.

**1.** $\frac{2}{5}$ $\qquad 5\overline{)2.0}$ quotient $0.4$ $\qquad$ So $\frac{2}{5} = 0.4$

**2.** $\frac{3}{4}$ $\qquad 4\overline{)3.00}$ quotient $0.75$ $\qquad$ So $\frac{3}{4} = 0.75$

**3.** $\frac{7}{9}$ $\qquad 9\overline{)7.000}$ quotient $0.777$ $\qquad$ So $\frac{7}{9} = 0.78$ (rounded off)

Here is a short list of common fractions converted to decimal equivalents. *It's a good idea to memorize these, since they are used so often.*

| Fraction | $\frac{1}{2}$ | $\frac{1}{3}$ | $\frac{2}{3}$ | $\frac{1}{4}$ | $\frac{3}{4}$ | $\frac{1}{5}$ | $\frac{2}{5}$ | $\frac{3}{5}$ | $\frac{4}{5}$ | $\frac{1}{8}$ |
|---|---|---|---|---|---|---|---|---|---|---|
| Decimal | 0.50 | 0.33* | 0.67* | 0.25 | 0.75 | 0.20 | 0.40 | 0.60 | 0.80 | 0.125 |

*These two decimals are rounded off from 0.3333 . . . and 0.6666 . . .

Common fractions with denominators of 10 are easily converted to decimal tenths.

*Example:* $\frac{3}{10} = 0.3$

**Exercises** Convert each fraction to a decimal to the nearest hundredth (2 places).

**1.** $\frac{3}{8}$ **2.** $\frac{5}{7}$ **3.** $\frac{9}{17}$ **4.** $\frac{21}{25}$ **5.** $\frac{17}{50}$

**Answers** **1.** 0.38 **2.** 0.71 **3.** 0.53 **4.** 0.84 **5.** 0.34

# Trial Test — Decimals

1. $0.71 + 0.032 + 1.5 =$
(1) 1.242
(2) 2.242
(3) 0.892
(4) 8.92
(5) 89.2

1. 1 2 3 4 5

2. $3.4 \times 51.8 =$
(1) 17,612
(2) 20.532
(3) 176.12
(4) 205.32
(5) 17.612

2. 1 2 3 4 5

3. $(0.7)^3 =$
(1) 3.43
(2) 343
(3) 0.343
(4) 0.21
(5) 2.1

3. 1 2 3 4 5

4. What decimal fraction is equivalent to $\frac{13}{25}$?
(1) 52
(2) 5.2
(3) 0.52
(4) 0.052
(5) 0.13

4. 1 2 3 4 5

5. What common fraction is the same as 0.103?
(1) $\frac{103}{100}$
(2) $\frac{103}{1,000}$
(3) $\frac{1}{103}$
(4) $\frac{10}{100}$
(5) $\frac{13}{100}$

5. 1 2 3 4 5

6. How much change from a $20 bill would you receive if you bought items costing $1.75, $3.39, and $6.21?
(1) $8.75
(2) $8.65
(3) $11.35
(4) $11.25
(5) $16.55

6. 1 2 3 4 5

7. Divide 5.38 by 100.
(1) 538
(2) 0.538
(3) 53.8
(4) 0.0538
(5) 0.00538

7. 1 2 3 4 5

8. At $2.85 per hour, what is the salary for a 40-hour week?
(1) $114
(2) $104
(3) $114.40
(4) $104.40
(5) None of the above

8. 1 2 3 4 5

9. Write as a decimal fraction: $2\frac{54}{1,000}$.
(1) 2.54
(2) 2.054
(3) 2.00054
(4) 0.254
(5) 0.00254

9. 1 2 3 4 5

10. Round off 453.0746 to the nearest hundredth.
(1) 453.07
(2) 543.08
(3) 453.1
(4) 4.53
(5) 453.075

10. 1 2 3 4 5

11. Find the sum: $\frac{1}{2} + 0.32 + \frac{1}{4}$
(1) 1.07
(2) 1.7
(3) 10.7
(4) 1.17
(5) 1.06

11. 1 2 3 4 5

12. $7.2 \times ? = 0.432$
(1) 3.1104
(2) 311.04
(3) 6
(4) 0.6
(5) 0.06

12. 1 2 3 4 5

## SOLUTIONS

1. **(2)** Remember to line up the decimal points in addition:

$$\begin{array}{r} 0.71 \\ 0.032 \\ \underline{1.5} \\ 2.242 \end{array}$$

2. **(3)** Multiply as if the numbers were whole numbers. Place the decimal point two places to the left to account for the two decimal places in the multipliers.

$$\begin{array}{r} 51.8 \\ \underline{\times \quad 3.4} \\ 207\ 2 \\ \underline{155\ 4\phantom{0}} \\ 176.12 \end{array}$$

3. **(3)** Remember $(0.7)^3$ means $0.7 \times 0.7 \times 0.7$.
$0.7 \times 0.7 = 0.49$ (two decimal places).
$0.49 \times 0.7 = 0.343$ (three decimal places).

4. **(3)** $\frac{13}{25}$ can be converted to a decimal by dividing 25 into 13.

$$\begin{array}{r} 0.52 \\ 25\overline{)13.00} \\ \underline{12\ 5\phantom{0}} \\ 50 \\ \underline{50} \\ 0 \end{array}$$

5. **(2)** 0.103 has three places. 0.1 means $\frac{1}{10}$, 0.10 means $\frac{10}{100}$, and 0.103 means $\frac{103}{1,000}$.

6. **(2)** To find the change, add the prices of the three items and subtract the result from $20.

$$\begin{array}{r} \$1.75 \\ 3.39 \\ \underline{6.21} \\ \$11.35 \end{array} \qquad \begin{array}{r} \$20.00 \\ \underline{-\ 11.35} \\ \$\ 8.65 \end{array}$$

7. **(4)** This can be done the hard way, by actually dividing 100 into 5.38. It is much easier to remember to move the decimal point two places to the left.
$5.38 \div 100 = 0.0538$

8. **(1)** At $2.85 per hour, the salary for 40 hours is just 40 × $2.85.

$$\begin{array}{r} \$2.85 \\ \underline{\times \quad 40} \\ \$114.00 \end{array}$$

9. **(2)** The word *and* indicates the decimal point, so we have 2 and something. $\frac{54}{1,000}$ is the same as 0.054, so the answer is 2.054.

**10.** **(1)** To round off 453.0746 to the nearest hundredth, notice that 7 is in the hundredth's place. The number to the right is 4, which is less than 5, so drop the remaining numbers. The rounded number is 453.07.

**11.** **(1)** To add this mixture of numbers, it is easier to convert the common fractions to decimals. $\frac{1}{2} = 0.5$ and $\frac{1}{4} = 0.25$. Then $\frac{1}{2} + 0.32 + \frac{1}{4}$ is the same as $0.5 + 0.32 + 0.25$, or 1.07.

**12.** **(5)** Think of this problem as "7.2 times something is 0.432." To find the "something," divide 7.2 into 0.432.

$$\begin{array}{r} 0.06 \\ 7.2_{\wedge}\overline{)\,0.4_{\wedge}32} \\ \underline{4\,32} \\ 0 \end{array}$$

## EVALUATION CHART

Multiply the number of correct answers by 100 and divide by the number of questions in the Trial Test to arrive at your score. ________

| Score | Rating |
|---|---|
| 90–100 | Excellent |
| 80–89 | Very Good |
| 70–79 | Average |
| 60–69 | Passing |
| 59 or Below | Very Poor |

# Chapter 4

# PERCENT

# 4.1 Meaning of "Percent"

So far, we have looked at two kinds of fractions, *common* and *decimal*. A third kind is perhaps most often used in business problems or in homemaking. This type of fraction is called percent.

> DEFINITION
>
> **Percent** simply means *hundredths*.

When a sign in a store reads "20% off," it means $\frac{20}{100}$ or $\frac{1}{5}$ off the regular price. When a bank advertises 5% interest, it means that money will increase in value $\frac{5}{100}$ of itself every year.

To calculate with percent, you must first change it to a fraction or a decimal. This is easy to do if you remember that percent means hundredths. Whenever the percent sign occurs, eliminate it and multiply by $\frac{1}{100}$.

*Examples:* **1.** 5% means $5 \times \frac{1}{100}$ or $\frac{5}{100}$

**2.** 20% means $20 \times \frac{1}{100}$ or $\frac{20}{100}$

**3.** 100% means $100 \times \frac{1}{100}$ or 1

# 4.2 Changing Percent to a Workable Form

Our first job is to change percent to a workable form, either a decimal or a common fraction. (Decimals are usually easier to work with, unless a "handy" common fraction comes up. A list of these "handy" equivalents will be included.)

Using the rule that percent means hundredths, 17% means $17 \times \frac{1}{100}$ or $\frac{17}{100}$. As a decimal, $\frac{17}{100}$ is the same as 0.17. 125% means $\frac{125}{100}$ or $1\frac{25}{100}$. This is the same as $1\frac{1}{4}$ or 1.25.

Sometimes a percent is given as a *mixed number* (a whole number and a fraction). Changing this to a decimal is still easy, but changing it to a common fraction takes a little care.

*Example:* Change $33\frac{1}{3}\%$ to a common fraction.

THINK: $33\frac{1}{3}\%$ means $33\frac{1}{3} \times \frac{1}{100}$.

First change $33\frac{1}{3}$ to an improper fraction: $\frac{100}{3}$.

So $33\frac{1}{3}\% = \frac{100}{3} \times \frac{1}{100}$, or $\frac{1}{3}$.

**Exercises** Try a few conversions from percent to common fractions.

1. 25%
2. $37\frac{1}{2}\%$
3. $66\frac{2}{3}\%$
4. $87\frac{1}{2}\%$
5. $83\frac{1}{3}\%$

**Answers**

1. $\frac{1}{4}$
2. $\frac{3}{8}$
3. $\frac{2}{3}$
4. $\frac{7}{8}$
5. $\frac{5}{6}$

Fractional percents must be handled with care. Take a look at this example to see how a percent is changed first to a fraction and then to a decimal.

$$\frac{1}{4}\% = \frac{1}{4} \times \frac{1}{100} \text{ or } \frac{1}{400}$$

To change this fraction to a decimal there are two ways:

1. Divide 400 into 1.

```
       0.0025
400 ) 1.000
        800
       2000
       2000
          0
```

*OR*

2. Think of the decimal equivalent for $\frac{1}{4}$. Remember that this is 0.25, so $\frac{1}{4}\%$ means 0.25%. Since 0.25% is $0.25 \times \frac{1}{100}$ (or $0.25 \times 0.01$), its decimal equivalent is 0.0025.

This shows us a simple rule to follow when we want to change a percent to a decimal.

> Since *percent* means *hundredths,* you can convert percent to a *decimal* by dropping the percent sign and moving the decimal point *two places to the left*.

For example, 5.5% becomes 0.055 and 0.3% is the same as 0.003.

**Exercises** Try these conversions, first to fractions and then to decimals.

| | Percent | Fraction | Decimal |
|---|---|---|---|
| **1.** | 60% | ? | ? |
| **2.** | 25% | ? | ? |
| **3.** | 3% | ? | ? |
| **4.** | 91% | ? | ? |
| **5.** | 150% | ? | ? |
| **6.** | 50% | ? | ? |
| **7.** | $66\frac{2}{3}\%$ | ? | ? |
| **8.** | $\frac{1}{2}\%$ | ? | ? |

**Answers**

| | Percent | Fraction | Decimal |
|---|---|---|---|
| **1.** | 60% | $\frac{3}{5}$ | 0.6 |
| **2.** | 25% | $\frac{1}{4}$ | 0.25 |
| **3.** | 3% | $\frac{3}{100}$ | 0.03 |
| **4.** | 91% | $\frac{91}{100}$ | 0.91 |
| **5.** | 150% | $1\frac{1}{2}$ or $\frac{3}{2}$ | 1.5 |
| **6.** | 50% | $\frac{1}{2}$ | 0.5 |
| **7.** | $66\frac{2}{3}\%$ | $\frac{2}{3}$ | $0.66\frac{2}{3}$ |
| **8.** | $\frac{1}{2}\%$ | $\frac{1}{200}$ | 0.005 |

# 4.3 Changing Decimals or Fractions to Percents

Our second job is to change a decimal or common fraction to a percent. For a decimal, it's easy: just reverse the steps above.

*Examples:* 0.17 means $\frac{17}{100}$ or 17%.

0.83 means $\frac{83}{100}$ or 83%.

0.5 is a different case. This means $\frac{5}{10}$, which cannot be directly changed to percent. Remember that 0.5 = 0.50, and this is $\frac{50}{100}$ or 50%.

The rule to follow is:

> To convert from *decimal* to *percent*, move the decimal point *two places to the right* and add the percent sign (%).

The reason for this is again that percent means hundredths. Thus 0.23, which means "23 hundredths," also means "23 percent."

> To change a *fraction* to a *percent, first change the fraction to a decimal,* and then follow the rule above.

*Example:* Change $\frac{7}{8}$ to a percent.

STEP 1 $\frac{7}{8}$ is changed to a decimal:

$$8 \overline{)7.000} = 0.875$$

STEP 2 To change 0.875 to a percent, move the decimal point two places to the right and add the percent sign:

87.5%

# 4.4 Arithmetic Problems Using Percent

Any arithmetic problem using percent falls into one of three categories:

1. Finding a number which is a given percent of another number.
2. Finding what percent one number is of another number.
3. Finding a number when a percent of it is known.

*Examples:* **1.** (First category) What is the sale price on a dress marked $49.95 with 20% off?

First, note how much is to be taken off. This is 20% of $49.95. The word "of" here means "times," so 20% of $49.95 means (0.20) × (49.95) or $9.99. This is the amount taken off the original price. Now subtract $9.99 from $49.95 and get $39.96. This is the sale price.

$$\begin{array}{r} \$49.95 \\ \times 0.20 \\ \hline \$9.9900 \end{array} \qquad \begin{array}{r} \$49.95 \\ -\ 9.99 \\ \hline \$39.96 \end{array}$$

**2.** (Second category) A baseball team won 17 out of 20 games played. What *percent* of its games did it win?

Here the key words are "out of." This tells us to compare the games won to the games played by *division*. The team won $\frac{17}{20}$ of its games. To convert to percent:

$$\begin{array}{r} 0.85 \\ 20\,\overline{)\,17.00} \\ \underline{16\,0} \\ 1\,00 \\ \underline{1\,00} \\ 0 \end{array}$$

So $\frac{17}{20} = 0.85$ or 85%.

**3.** (Third category. This case — finding a whole number when a fraction or percent of it is known — requires special handling). Suppose $\frac{1}{4}$ pound of salami costs 45¢. What does a pound cost?

It's easy to see that we multiply $\frac{1}{4}$ by 4 to get a whole pound, so 4 × 45¢ is $1.80.

THINK: $\frac{1}{4} \times ? = 45$. To find the missing number, we multiply both sides by 4.

Let's try the same method on a different problem. A bank pays 5% interest on savings. If the interest for one year is $135, how big is the *original* savings deposit?

First, let's write the problem in a different way.

5% of some number is $135.

The word "of" means *times* and "is" means *equals*. We could use a question mark to stand for the missing number. So we rewrite:

$$0.05 \times ? = 135 \text{ or } \frac{5}{100} \times ? = 135$$

Then, if we multiply $135 by 100 and divide by 5, we'll know the missing number.

$$\frac{135 \times 100}{5} = \$2{,}700.$$

Think about the following:

5 times a number is 35. What is the number? Surely the number is 7. But we could write the statement $5 \times ? = 35$.

This same idea applies to $0.05 \times ? = 135$.

Divide $\frac{135}{0.05}$ to get $2,700

Now the rule:

> To find a number when a *percent* of it is known, *change the percent to a decimal* and *divide* into the part which is known.

*Any problem involving percent fits into one of these three cases.* Your problem is to decide which it is, and which rule to follow.

**Exercises**

1. Find 10% of $50.
2. What percent is 12 feet of 15 feet?
3. $20 is what percent of $25?
4. 60% of what number is 72?
5. 35% of $200 is how much?
6. 12 is what percent of 20?
7. 12% of 20 is ?
8. 10% of what number is 71?
9. What percent of 50 is 15?
10. 16 is what percent of 80?

**Answers**

**1.** $5 **3.** 80% **5.** $70 **7.** 2.4 **9.** 30%

**2.** 80% **4.** 120 **6.** 60% **8.** 710 **10.** 20%

It's a good idea to memorize a few of the most common percents and their equal fractional values. Study the table below.

| $\frac{1}{2}$ | $\frac{1}{3}$ | $\frac{2}{3}$ | $\frac{1}{4}$ | $\frac{3}{4}$ | $\frac{1}{5}$ | $\frac{2}{5}$ | $\frac{3}{5}$ | $\frac{4}{5}$ |
|---|---|---|---|---|---|---|---|---|
| 50% | $33\frac{1}{3}$% | $66\frac{2}{3}$% | 25% | 75% | 20% | 40% | 60% | 80% |
| $\frac{1}{8}$ | $\frac{3}{8}$ | $\frac{5}{8}$ | $\frac{7}{8}$ | $\frac{1}{10}$ | $\frac{3}{10}$ | $\frac{7}{10}$ | $\frac{9}{10}$ | |
| $12\frac{1}{2}$% | $37\frac{1}{2}$% | $62\frac{1}{2}$% | $87\frac{1}{2}$% | 10% | 30% | 70% | 90% | |

Notice that the fractions $\frac{2}{8}, \frac{4}{8}, \frac{6}{8}, \frac{2}{10}, \frac{4}{10}, \frac{8}{10}$, are missing from the table above. This is so because all these fractions reduce to lower terms which appear in the other columns $(\frac{1}{4}, \frac{1}{2}, \frac{3}{4}, \frac{1}{5}, \frac{2}{5}, \frac{4}{5})$.

Now here are some more difficult examples. Notice how they fit into the three cases of percent problems.

*Examples:* **1.** A real estate agent receives a 6% commission on the sale of a house. This amount is deducted from the selling price. How much will the seller receive when a house is sold for $12,000?

There are some key words here. "Commission" is an amount paid to the agent for the service he or she performs. "Selling price" is just what it seems: what the buyer must pay. What are the steps to solve the problem?

STEP 1 Find 6% of $12,000.
$0.06 \times 12{,}000 = \$720$

STEP 2 Subtract $720 from $12,000.
$12{,}000 - 720 = \$11{,}280$

**2.** Over a 5-year period, the population of a town rose from 10,000 to 12,400. Find the percent of increase in population.

This fits into case 2, finding what percent one number is of another. How much increase was there?

$$12{,}400 - 10{,}000 = 2{,}400$$

Now compare this increase to the *original* population to find the percent of increase.

$$\frac{2{,}400}{10{,}000} = 0.2400 \text{ or } 24\% \text{ increase.}$$

**3.** Simple interest is an often used application of percent. A formula for interest is:

$$i = p \times r \times t$$

The $i$ means *interest* or *income*. The $p$ stands for *principal* or money invested. The $r$ is the *rate* of interest in percent. The $t$ means *time* in years.

How much simple interest will there be on $500 for 2 years at 5%?

Use the formula: $i = p \times r \times t$
$i = 500 \times 0.05 \times 2$
or $\frac{500}{1} \times \frac{5}{100} \times \frac{2}{1}$
$i = \$50$

# Trial Test — Percent

**1.** What is 25% of 16.4?
(1) 4.1
(2) 41
(3) 410
(4) 6.56
(5) 656

**1.** 1 2 3 4 5
|| || || || ||

2. On a test of 25 items, 18 were answered correctly. What percent were answered correctly?
(1) 0.72%
(2) 72%
(3) 42%
(4) 7.2%
(5) 4.2%

2. 1 2 3 4 5

3. If 8% of a number is 7.2, what is the number?
(1) 0.576
(2) 57.6
(3) 9
(4) 90
(5) 900

3. 1 2 3 4 5

4. Express $3\frac{1}{2}\%$ as a common fraction.
(1) $\frac{7}{2}$
(2) $\frac{7}{200}$
(3) 0.035
(4) 0.35
(5) $\frac{7}{20}$

4. 1 2 3 4 5

5. On a bank account of $1,600, how much yearly interest will be paid if the rate is 5%?
(1) $8
(2) $80
(3) $800
(4) $32
(5) $320

5. 1 2 3 4 5

6. If a color television set regularly sold at $595 is marked down 20%, what is the sale price?
(1) $119
(2) $476
(3) $505
(4) $583.20
(5) $575

6. 1 2 3 4 5

7. Express 0.3% as a decimal.
(1) 0.3
(2) 3
(3) 0.003
(4) 0.0003
(5) 0.03

7. 1 2 3 4 5

8. Change $\frac{3}{200}$ to a percent.
(1) 1.5%
(2) 0.15%
(3) 0.015%
(4) $66\frac{2}{3}\%$
(5) $6\frac{2}{3}\%$

8. 1 2 3 4 5

9. On his diet, John's weight dropped from 200 to 178 pounds. What was the percent of decrease in his weight?
(1) 12.4%
(2) 1.1%
(3) 1.24%
(4) 11%
(5) 22%

9. 1 2 3 4 5

10. A salesperson works on a 10% straight commission. If her sales for one week were $2,480, how much did she earn?
(1) $24.80
(2) $10
(3) $248
(4) $2.48
(5) $100

10. 1 2 3 4 5

11. A house is assessed at 20% of its real value. If it is assessed at $3,800, what is its real value?
(1) $760
(2) $7,600
(3) $19,000
(4) $1,900
(5) $3,040

11. 1 2 3 4 5

12. When a coin was tossed, it came up *heads* 12 times and *tails* 8 times. What percent of the time did it come up tails?
(1) 40%
(2) $66\frac{2}{3}\%$
(3) 8%
(4) 4%
(5) 20%

12. 1 2 3 4 5

## SOLUTIONS

1. **(1)** This is case 1. Remember "of" means times. 25% of 16.4 is the same as 0.25 × 16.4 or $\frac{1}{4}$ × 16.4. Either way, the answer is 4.1.

2. **(2)** This is case 2, finding what percent one number is of another. 18 out of 25 means $\frac{18}{25}$. As a decimal, this is 0.72 and as a percent, move the decimal point 2 places to the right to get 72%.

3. **(4)** This is case 3, finding the number when the part is known. Remember: divide 0.08 into 7.2.

$$0.08_\wedge \overline{)\,7.20_\wedge} = 90.$$

4. **(2)** First, change $3\frac{1}{2}$ to an improper fraction. $3\frac{1}{2}\% = \frac{7}{2}\%$. Now remember that percent means "hundredths," so $\frac{7}{2}\%$ is the same as $\frac{7}{2} \times \frac{1}{100}$ or $\frac{7}{200}$.

5. **(2)** Remember the formula $i = p \times r \times t$. In this case $p$ = 1,600, $r$ is 5% and $t$ is 1 year.

$i$ = 1,600 × 0.05 × 1 or $80.

6. **(2)** First find the amount of decrease, 20% of $595.

0.20 × 595 = $119

Then subtract $119 from $595 to get $476, the net price.

7. **(3)** This is an easy one. Just remember to move the decimal point 2 places to the left to change a percent to a decimal.

0.3% = 0.003

8. **(1)** First change $\frac{3}{200}$ to a decimal, dividing 200 into 3.

$$200 \overline{)\,3.000} = 0.015$$

Then move the decimal point 2 places to the right.

$0.01_\wedge5 = 1.5\%$

9. **(4)** This is case 2 again, finding what percent one number is of another. What two numbers are being compared? The actual decrease is 22 pounds compared to the orig-

inal weight, 200 pounds. $\frac{22}{200}$ is changed to a decimal:

$$200\overline{)22.00}\quad = 0.11$$

```
       0.11
200 ) 22.00
      20 0
       2 00
       2 00
          0
```

and 0.11 = 11%.

**10.** **(3)** This is 10% of $2,480, or (0.10) × (2,480) = $248.

**11.** **(3)** Rewriting the question: "3,800 is 20% of what number?" This is case 3, finding the number when the part (percent) is known. Divide 0.20 into 3,800.

```
         1900 0.
0.2^ ) 3800.0^
```

So the real value is $19,000.

**12.** **(1)** This time, compare the number of tails, 8, to the *total* number of tosses, 20.

$$\frac{8}{20} = 0.40 \text{ or } 40\%.$$

## EVALUATION CHART

Multiply the number of correct answers by 100 and divide by the number of questions in the Trial Test to arrive at your score. ______

| Score | Rating |
|---|---|
| 90–100 | Excellent |
| 80–89 | Very Good |
| 70–79 | Average |
| 60–69 | Passing |
| 59 or Below | Very Poor |

# Chapter 5

# BEGINNING ALGEBRA

As mathematicians searched to simplify the solution of problems, over the years they developed a new tool to assist them. This tool came to be known as <u>algebra</u>. Essentially, algebra is just a way to reduce a problem to a small set of symbols. When we can do this, the solution is usually easy to find. To use algebra we must deal with "unknowns" or "variables." Let's take a look at these letters which stand for numbers.

# 5.1 Literal Numbers

Here is a new way to think of numbers. We probably all understand that if a room is 9′ by 12′, it has an area of 9 × 12 or 108 square feet; also, if the room were 12′ by 18′, its area would be 12 × 18 or 216 square feet. Is there a logical way to talk about the area of any room? Yes. We may say that for any room the area is the product of the length and the width.

Think about that statement. It's a long one, but it can be boiled down to "Area = length × width." We can do still better by using initials to stand for the numbers. The whole statement becomes:

$$A = l \times w$$

As we look at $A = l \times w$, we think: "$A$ means *area,* $l$ means *length*, and $w$ means *width*." So, in this case, we see a new way of talking about not just one number, but a collection of numbers.

The letters $A$, $l$, and $w$ are called <u>*literal numbers*</u>, because they are *letters that stand for numbers.* So a *literal number* is a letter that takes the place of a number. Another word for these letters is *variables*. This is logical because the $A$, $l$, and $w$ are allowed to have different values; they *vary* or change.

# 5.2 Operations

The same operations of arithmetic are used in algebra: addition, subtraction, multiplication, and division. Suppose you are asked to find the sum of two numbers. If the numbers are, say, 5 and 3, the answer is 5 + 3 or 8. But, if the numbers are not known, call them $x$ and $y$. Then $x$ and $y$ stand for *any* two numbers, so you can write the sum as "$x + y$."

The same idea holds for all the operations:

| | |
|---|---|
| The sum of two numbers $x$ and $y$: | $x + y$ |
| The difference of two numbers $x$ and $y$: | $x - y$ |
| The product of two numbers $x$ and $y$: | $(x) \times (y)$ or $x \cdot y$ or $xy$* |
| The quotient of two numbers $x$ and $y$: | $x \div y$ or $\frac{x}{y}$ where $y \neq 0$ |

*To show multiplication of two literal numbers, the "times" sign is usually omitted. Both $x \cdot y$ and $xy$ mean "$x$ times $y$."

Let's see how to use symbols to translate simple statements.

*Examples:* Represent the following with symbols.

1. The sum of $a$ and $b$: $a + b$ or $b + a$
2. The product of $m$ and $n$: $m \times n$ or $m \cdot n$ or $mn$ or $nm$
3. The quotient of $p$ divided by 3: $p \div 3$ or $\frac{p}{3}$
4. The quotient of 3 divided by $p$: $3 \div p$ or $\frac{3}{p}$
5. Twice a number: $2 \cdot x$ or $2x$
6. A number decreased by 7: $n - 7$
7. 7 decreased by a number: $7 - n$
8. A number, $x$, increased by another number, $y$: $x + y$
9. Twice one number, $x$, decreased by three times another, $y$: $2x - 3y$.

**Exercises**

Now try a few translations on your own. Before you approach a problem using algebra, you must first be able to translate from words to symbols such as $x$ and $y$.

1. 4 more than $y$
2. 10 less than $y$
3. The difference between 8 and $x$
4. $x$ subtracted from 8
5. The difference between $x$ and 8
6. 8 subtracted from $x$
7. $h$ increased by $k$
8. 7 added to $n$
9. $p$ decreased by $q$
10. The product of $x$ and twice $y$

**Answers**

1. $y + 4$ or $4 + y$
2. $y - 10$, *not* $10 - y$
3. and 4. $8 - x$, *not* $x - 8$
5. and 6. $x - 8$, *not* $8 - x$
7. $h + k$ or $k + h$
8. $n + 7$ or $7 + n$
9. $p - q$, *not* $q - p$
10. $2xy$ or $x2y$ or $xy2$. By custom, we prefer to write the "2" first, for ease of handling.

# 5.3 Equations

An *equation* is a statement that two quantities are equal.

*For example:* $3 + 7 = 10$ $\quad 7 \cdot 8 = 56$

$10 - 7 = 3$ $\quad 18 \div 2 = 9$

For our purposes in algebra, equations will always contain variables or unknowns. In every equation, our objective will be to find a replacement number for the variable that will make the equation a true statement.

Think about $x + 5 = 7$. This is a statement which means that some number plus 5 is equal to 7. Ask yourself the question, "What can I use for $x$ to make the statement true?" When $x$ is replaced by 2, the statement is true. So we say that 2 is the *solution* to the equation.

> DEFINITION
>
> The **solution** to an equation is the number(s) which, when used as the value of the variable(s), makes the statement true.

Think about the following equations.

1. $x - 3 = 11$
2. $3 \cdot x = 21$
3. $x \div 5 = 4$

These statements would be read:

1. A number minus 3 is equal to 11.
2. 3 times a number is equal to 21.
3. A number divided by 5 is equal to 4.

Use arithmetic to find the unknown number.

1. The replacement for $x$ is 14, because $14 - 3 = 11$.
2. The replacement for $x$ is 7, because $3 \cdot 7 = 21$.
3. The replacement for $x$ is 20, because $20 \div 5 = 4$.

Think about each equation as a balance scale. For example: $10 = 10$. If one side is changed in some way, the other side must be changed in exactly the same way, if we are to keep the two sides equal.

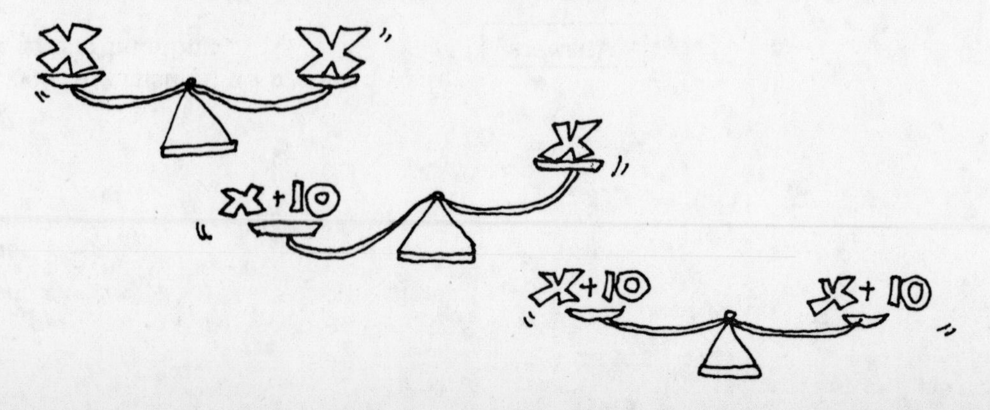

If we start with $10 = 10$, and change both sides of the equation by the same amount, the equation will still be balanced.

$$10 + 5 = 10 + 5$$
$$15 = 15$$
$$10 - 3 = 10 - 3$$
$$7 = 7$$

$$10 \cdot 7 = 10 \cdot 7$$
$$70 = 70$$
$$10 \div 2 = 10 \div 2$$
$$5 = 5$$

What does all this have to do with solving equations? Let us go back to $x + 5 = 7$. This means "some number increased by 5 equals 7." To find the solution, the final form of the equation must read "$x$ = (some number)." The variable must be *isolated* on one side of the equation. Since $x$ has been increased by 5 in our present equation, we must decrease the left side of the equation by 5 in order to find $x$.

Then, thinking about the balance scale, we must decrease the right side by the *same amount*.

$$x + 5 = 7$$
$$x + 5 - 5 = 7 - 5$$
$$x = 2$$

The same idea will help you solve every simple equation.

*Examples:* **1.** $x - 3 = 7$

This means "a number, decreased by 3, is equal to 7." To find the number, we must *increase* each side by 3.

$$(x - 3) + 3 = 7 + 3$$

This procedure isolates $x$, so we have found the solution:

$$x = 10$$

**2.** $5x = 30$

This means "5 times a number is equal to 30." To find the number, we must *divide* each side by 5.

$$\frac{5x}{5} = \frac{30}{5}$$

We are left with:

$$x = 6$$

**3.** $x \div 2 = 6$

This means "a number divided by 2 is equal to 6." Since the number has been *divided* by 2, each side now must be *multiplied* by 2.

$$2\left(\frac{x}{2}\right) = 2(6)$$
$$x = 12$$

What common idea appears in each of the examples? It seems that each time you see an operation: $+$, $-$, $\times$, $\div$, meaning addition, subtraction, multiplication, or division, it must be *undone* by using the *inverse* or *opposite* operation.

TO SOLVE AN EQUATION:

1. What operation do you see?
2. What is the *inverse* operation?
3. Use the inverse operation on *each* side in order to isolate the variable. *Be sure to use the same operation on each side of the equation,* or you will destroy the balance and the two sides of your solution equation will not be equal!

*Examples:* **1.** $x + 7 = 50$

You see *addition,* so *subtract* 7 from each side of the equation to remove 7 from the left side.

$$x + 7 - 7 = 50 - 7$$

7 from 7 is 0, and 7 from 50 is 43. We are left with:

$$x = 43$$

$x = 43$ is the solution.

**2.** $0.05x = 4$

$0.05x$ means "0.05 *times x*," so *divide* each side by 0.05 to isolate $x$.

$$\frac{0.05x}{0.05} = \frac{4}{0.05}$$

$0.05 \div 0.05 = 1$. $4 \div 0.05 = 80$. So we are left with:

$$x = 80$$

**3.** $x - 3 = 4$

You see *subtraction,* so *add* 3 to each side.

$$x - 3 + 3 = 4 + 3$$

We are left with:

$$x = 7$$

**4.** $\frac{x}{2} = 7$

Here you see *division* so *multiply* each side by 2.

$$2\left(\frac{x}{2}\right) = 2(7)$$

$$x = 14$$

**Exercises** Find the value of $n$ that will make each of these equations true.

**1.** $n - 7 = 30$
**2.** $n - 6 = 8$
**3.** $5n = 25$
**4.** $\frac{n}{4} = 6$
**5.** $n \div 2 = 8$
**6.** $6n = 42$
**7.** $n + 20 = 30$
**8.** $n - 4 = 12$

9. $n + a = b$

10. $n - a = b$

11. $an = b$

12. $\dfrac{n}{a} = b$

Answers

1. $n = 37$
2. $n = 14$
3. $n = 5$
4. $n = 24$
5. $n = 16$
6. $n = 7$
7. $n = 10$
8. $n = 16$
9. $n = b - a$, *not* $a - b$
10. $n = a + b$ or $b + a$
11. $n = \dfrac{b}{a}$ or $b \div a$, *not* $\dfrac{a}{b}$ or $a \div b$; remember, too, that in exercises 11 and 12, $a$ cannot = zero.
12. $n = ab$ or $n = ba$

# 5.4 Signed or Directed Numbers

When the temperature outside is $-10°$, we know it's very cold. Just how cold is it? What meaning does $-10°$ have? Clearly it means 10 degrees below zero. Now we're ready to investigate a new kind of number: numbers which have a *direction* or sign.

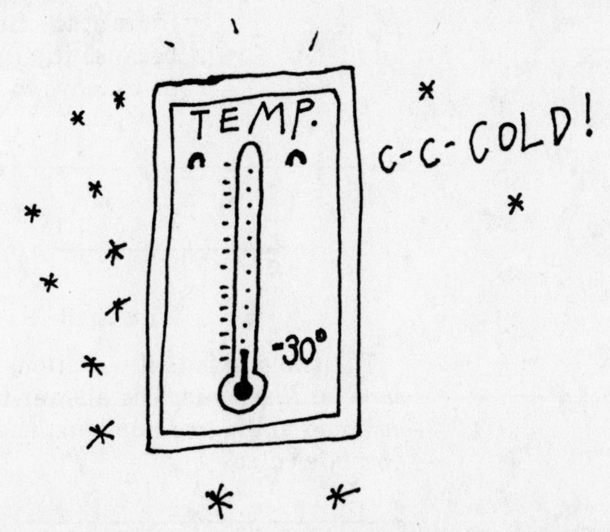


A good way to understand signed numbers is to keep in mind a number line similar to the one pictured below. You might think of it as a thermometer in a horizontal position.

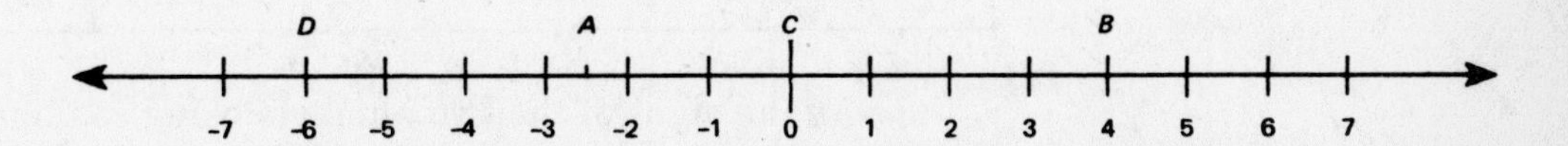


The numbers to the right of zero generally are shown without the plus sign. So, for example, 3 means the same as $+3$. (For clarity, we will use the plus sign in equations throughout the next section.) Negative numbers are *always* shown with a minus sign. On the number line above, points $A$, $B$, $C$, and $D$ correspond to the numbers $-2\frac{1}{2}$, $+4$, 0, and $-6$, respectively.

REMEMBER: 0 (zero) has no sign. It is neither positive nor negative.

# 5.5 Operations with Signed Numbers

## ADDITION

To add signed numbers, it is important to keep the number line in mind.

*Examples:* **1.** $(+3) + (+5)$

Start at the first number, $+3$, and travel 5 units to the right.

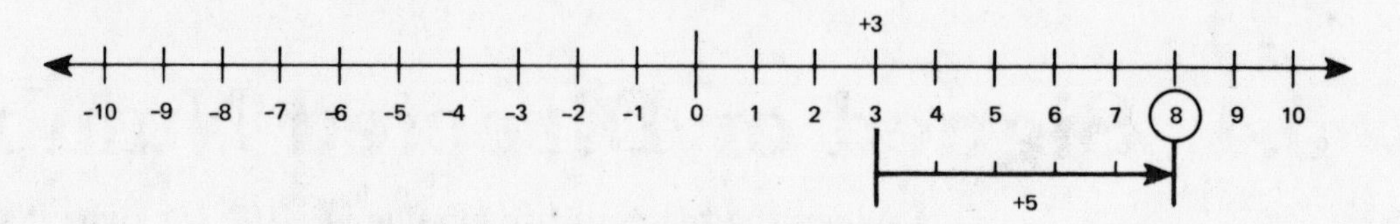


Thus $(+3) + (+5) = +8$

**2.** $(-3) + (-5)$

Start at the first number, $-3$, and travel 5 units to the left, because the minus sign with the second number, $-5$, tells us to move to the left.

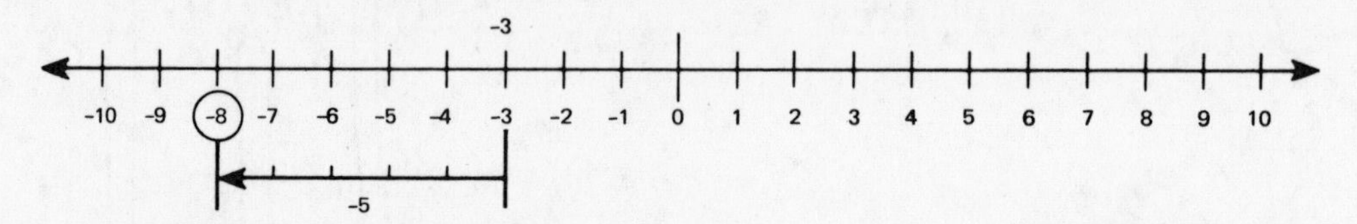


Thus $(-3) + (-5) = -8$

The two previous illustrations give us an idea for adding numbers with the *same* or *like* signs. The answer will, of course, have the same sign as the one common to the two numbers, and a numerical value equal to the sum of their *absolute values*.

> DEFINITION
>
> **Absolute value** is the unsigned or undirected value of a number. The absolute value of $-3$ is 3, and the absolute value of $+5$ is 5.

Suppose the signs of the two numbers being combined are *different*.

*Examples:* **1.** $(+3) + (-5)$

Start, as before, with the point on the number line corresponding to the first number, $+3$, and travel 5 units to the *left*, because the minus sign on the second number, $-5$, tells us to move to the left from our starting point, $+3$.

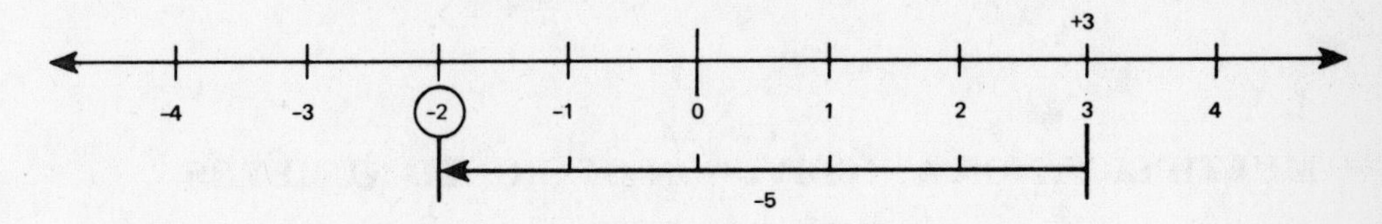


Thus $(+3) + (-5) = -2$

One interpretation of the example above might be to suppose that the thermometer shows a reading of 3° and then the temperature drops 5°. What will be the reading after this drop? We see that our answer is, as we have shown, 2° below zero, or $-2°$.

**2.** $(-3) + (+5)$

We start as before with our first number, $-3$, and move 5 units to the *right*, because our second number is positive this time.

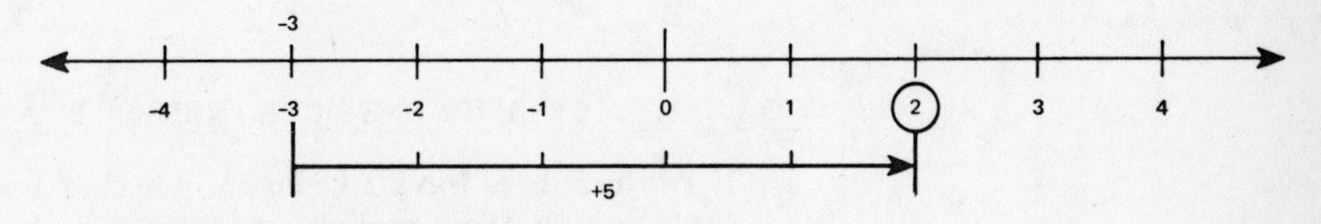


Thus $(-3) + (+5) = +2$

TO ADD SIGNED NUMBERS:

1. If their signs are *alike, add* their absolute values, and *keep the common sign.*
2. If the signs are *unlike*, or *different,* find the *difference* between the absolute values of the two numbers, and *keep the sign of the number with the greater numerical value.*

## SUBTRACTION

Since subtraction is the *opposite* of addition, the subtraction of signed numbers is just *adding the opposite.* $(+5) - (+3)$ is the same as $(+5) + (-3)$. They both give us 2.

TO SUBTRACT SIGNED NUMBERS:

1. Change the sign of the *second* number.
2. Use the rule for *adding.*

*Examples:* **1.** $(+5) - (-3)$ $= (+5) + (+3)$ $= +8$ **2.** $(-5) - (-3)$ $= (-5) + (+3)$ $= -2$ **3.** $(-5) - (+3)$ $= (-5) + (-3)$ $= -8$

## MULTIPLICATION AND DIVISION OF SIGNED NUMBERS

Before establishing a rule for multiplying signed numbers, let's take time out for an illustration. Think of money saved as positive (+), money spent as negative (−), time future as positive (+) and time past as negative (−). Thus, if we save money at a rate of $5 per week (+5), then three weeks in the future (+3), we will be $15 better off than we are today (+15). So $(+3) \cdot (+5) = +15$.

Now if we *spend* money at a rate of $5 per week (−5), three weeks from now (+3), we'll be *worse off* by $15. So, $(-5) \cdot (+3) = -15$.

If we have been spending money at a rate of $5 per week (−5), three weeks ago (−3), we were better off than today by $15 (+15). Thus the product of −5 and −3 is +15. From these examples, we can draw the following conclusions:

TO MULTIPLY OR DIVIDE ONE SIGNED NUMBER BY ANOTHER:

1. If two numbers have *like signs,* whether both are positive or both are negative, their *product or quotient is always positive.*
2. If they have *different signs,* their *product or quotient is negative.*

*Examples:*
**1.** $(+5)(+3) = +15$
**2.** $(-5)(-7) = +35$
**3.** $(+15) \div (-3) = -5$
**4.** $(-24) \div (-6) = +4$
**5.** $(+6)(-5) = -30$
**6.** $(-8)(+5) = -40$
**7.** $(-0.5) \div (-1) = +0.5$
**8.** $(-0.9) \div (-0.1) = +9$

In finding the product of *more than two* numbers, you must remember that we operate on *only two* numbers at a time. So we start by finding the product of *any two* and continue until we have used all of the factors.

*Example:*
$$(+2)(-5)(+3)(-3) = (-10)(+3)(-3) = (-30)(-3) = +90$$
or:
$$(+2)(-5)(+3)(-3) = (-10)(-9) = +90$$

An easy way to find the sign of the product is to count the number of minus signs. An even number of minus signs will produce a positive product. Why? *Because the product of two negative numbers is positive.*

*Examples:* **1.** $(-2)(-3)(-4)(-5) = (+6)(+20) = +120$

**2.** $(-3)^3$ means $(-3)(-3)(-3)$. An odd number of minus signs tells us the product is negative: $-27$.

**Exercises**

1. $(+3) + (-7)$
2. $(+12) + (-13) + (+1)$
3. $(-5)(-4)(-2)$
4. $(-3)^4$
5. $(-12) \div (-2)$
6. $(+6) - (-6)$
7. $(-4)(+3)(-2)(-1)$
8. $(-4) + (+3) + (-2) + (-1)$
9. $(-6)(-5) \div (-2)$
10. $(+7) + (-2)(-3) - (+7)$

**Answers**

1. $-4$ 2. 0 3. $-40$ 4. $+81$ 5. $+6$ 6. $+12$ 7. $-24$ 8. $-4$ 9. $-15$ 10. $+6$

# 5.6 Using Signed Numbers in Equations

With a knowledge of signed numbers we can solve more complex equations and problems.

*Examples:* **1.** $3x + 7 = -11$

STEP 1 Subtract 7 from each side to remove 7 from the left side.

$3x + 7 - 7 = -11 - 7$

We are left with:

$$3x = -18$$

STEP 2 Now divide each side by 3 to find $x$.

$$\frac{3x}{3} = \frac{-18}{3}$$

$$x = -6$$

CHECK Substitute 6 for $x$ in the original equation.

$3(-6) + 7 \stackrel{?}{=} -11$ ($\stackrel{?}{=}$ means we're not yet sure if $-6$ balances the equation)

$-18 + 7 \stackrel{?}{=} -11$ The equation balances out, so
$-11 = -11$ $-6$ is the correct solution.

**2.** $-7x = 24 - x$

STEP 1 Add $x$ to each side to remove $x$ from the right side.

$$-7x + x = 24 - x + x$$
$$-6x = 24$$

STEP 2 Now divide each side by $-6$ to find $x$.

$$\frac{-6x}{-6} = \frac{24}{-6}$$
$$x = -4$$

CHECK Substitute $-4$ for $x$ in the original equation.

$$-7(-4) \stackrel{?}{=} 24 - (-4)$$
$$28 \stackrel{?}{=} 24 + 4$$
$$28 = 28$$

**3.** $8x - 2 - 5x - 8 = 0$

STEP 1 Regroup the terms.

$$(8x - 5x) + (-2 - 8) = 0$$

STEP 2 Do the operations inside the parentheses.

$$3x - 10 = 0$$

STEP 3 Now add 10 to each side.

$$3x - 10 + 10 = 0 + 10$$
$$3x = 10$$

STEP 4 Divide each side by 3 to find $x$.

$$\frac{3x}{3} = \frac{10}{3}$$
$$x = \frac{10}{3}$$

CHECK Substitute $\frac{10}{3}$ for $x$ in the original equation.

$$8\left(\frac{10}{3}\right) - 2 - 5\left(\frac{10}{3}\right) - 8 \stackrel{?}{=} 0$$
$$\frac{80}{3} - 2 - \frac{50}{3} - 8 \stackrel{?}{=} 0$$

Now combine the fractions and combine the whole numbers.

$$\frac{30}{3} - 10 \stackrel{?}{=} 0$$

Change $\frac{30}{3}$ to a whole number.

$$10 - 10 \stackrel{?}{=} 0$$
$$0 = 0$$

Exercises

**1.** $3x = 7 - 4x$

**2.** $6 - 3x + 12 = 0$

**3.** $-7 + 5x = 10x - 2$

**4.** $6y - 2 + 2y = 14$

**5.** $5x + 12 - 3x = -2x - 13 - x$

**Answers**

1. $x = 1$
2. $x = 6$
3. $x = -1$
4. $y = 2$
5. $x = -5$

# 5.7 Algebraic Expressions

An <u>*algebraic expression*</u> is just a collection of numbers and variables. $3x + 4y$ is an algebraic expression meaning "3 times a number $x$ plus 4 times another number $y$." $2x^3$ is an algebraic expression meaning "2 times $x$ times $x$ times $x$." It's important to use the correct names for different algebraic expressions.

DEFINITIONS

**Term:** An algebraic expression with numbers and variables connected only by the operations of multiplication and division.

*Examples:* $3x$, $2xy$, $-7a^2$, $4abc$, $\frac{3}{a}$, $\frac{a^2b}{c^2}$

Since terms are formed using only multiplication and division, plus and minus signs separate terms and make expressions with more than one term.

**Monomial:** An algebraic expression with one, and *only one*, term.

*Examples:* $2abfg$, $x^2y^4$, $(2)(3)xyz$, $\frac{xyz}{3}$

**Binomial:** An algebraic expression with two, and *only two*, terms separated by either plus or minus signs.

*Examples:* $2x - 3y$, $x^2 - 4y^2$, $ab + 4a^2$, $abcd + 5d$

DEFINITIONS *(cont.)*

**Trinomial:** An algebraic expression with three, and *only three,* terms, separated by either plus or minus signs.

*Examples:* $x^2 + 2x - 7$, $3x - 4y + z$

**Polynomial:** Any algebraic expression with *more* than *one* term. *Polynomial* is a general name for binomials as well as trinomials, but *not* for monomials.

*Examples:* $2x + 3y$, $abc - 7cd - 8ce$, $2x - 3y - 4z + 4a - 2b$

**Similar terms:** Terms are *similar* or *like* when they have the *same literal factors* raised to the *same powers.*

*Examples:* $3x$ and $2x$ are like terms; $3x$ and $2x^2$ are *not*.
$3x^2$ and $-x^2$ are like terms; $3x^2$ and $x$ are *not*.
$x^2z$ and $-7x^2z$ are like terms; $x^2z$ and $-7xz$ are *not*.

# 5.8 Adding Algebraic Expressions

If you were to add 3′5″ to 7′4″, the answer would be 10′9″. Notice that we add feet to feet and inches to inches. This same idea carries over to algebraic expressions.

When adding algebraic expressions, remember that only *like* terms can be combined.

*Example:* Simplify $3x + 2y - 4z + 2x - 5y$.
You can combine $3x$ and $2x$ into the single term $5x$.
You can combine $2y$ and $-5y$ into the single term $-3y$.
$5x - 3y - 4z$ is the *sum*.

# 5.9 Multiplying Algebraic Expressions

Let's begin with single-term expressions.

TO MULTIPLY ALGEBRAIC EXPRESSIONS:

First, *multiply the numerical factors,* then *multiply the literal factors* (letter variables). When multiplying one power of $x$ by another power of $x$, simply *add the exponents.*

*Examples:* $(2x^2)(3x^5)$
$(2)(3) = 6$ and $(x^2)(x^5) = x^7$ $(xx \cdot xxxxx = xxxxxxx$, or $x^7)$
$6x^7$ is the *product*.

To multiply more complex expressions, remember the distributive law: $a(b + c) = ab + ac$.

*Examples:* **1.** $2(x - 3y) = 2x - 6y$
**2.** $3x^2(5a + 2b) = 15ax^2 + 6bx^2$
**3.** $ab(a + b) = a^2b + ab^2$
**4.** $x^2y(3x - 5y) = 3x^3y - 5x^2y^2$

And now a more difficult example. See how the distributive law works here:

**5.** $(a + 2)(2a - 3)$

HINT: Think of $(a + 2)$ as the multiplier.

$$(a + 2)(2a) + (a + 2)(-3) = 2a^2 + 4a - 3a - 6$$
$$= 2a^2 + a - 6$$

An easier way to perform this multiplication is to do four *separate* multiplications. Then the procedure looks like the ordinary multiplication of arithmetic:

$$\begin{array}{r} 2a - 3 \\ a + 2 \\ \hline 4a - 6 \\ 2a^2 - 3a \quad \\ \hline 2a^2 + a - 6 \end{array}$$

$4a - 6$: Multiply $(2a - 3)$ by 2.
$2a^2 - 3a$: Next multiply $(2a - 3)$ by $a$.
$2a^2 + a - 6$: Add partial products as you do in arithmetic.

# 5.10 Division of One Algebraic Expression by Another

TO DIVIDE ALGEBRAIC EXPRESSIONS:

First *divide the numerical factors,* and then *divide the literal factors.* When dividing one power of $x$ by another power of $x$, simply *subtract the exponents.*

Remember the meaning of exponents, and be careful not to make the mistake of thinking that $\frac{a^2}{b^2} = \frac{a}{b}$. The literal factors must be the same before you divide. Thus, $x^5 \div x^3$ means the same as $\frac{x^5}{x^3}$ or $\frac{x \cdot x \cdot x \cdot x \cdot x}{x \cdot x \cdot x} = x \cdot x$ or $x^2$.

*Examples:* **1.** $\frac{8x^3}{4x} = \frac{\overset{2}{\cancel{8}}}{\underset{1}{\cancel{4}}} \cdot \frac{\overset{x^2}{\cancel{x^3}}}{\underset{1}{\cancel{x}}} = 2 \cdot x^2$ or $2x^2$

2. $$\frac{15a^3b^2c}{5abc} = \frac{\overset{3}{\cancel{15}}}{\underset{1}{\cancel{5}}} \cdot \frac{\overset{a^2}{\cancel{a^3}}}{\underset{1}{\cancel{a}}} \cdot \frac{\overset{b}{\cancel{b^2}}}{\underset{1}{\cancel{b}}} \cdot \frac{\cancel{c}}{\underset{1}{\cancel{c}}} = 3a^2b$$

3. $$\frac{24x^3 - 16x^2 - 8x}{8x}$$

This example is more difficult. *Each* term must be divided by $8x$.

$$\frac{24x^3}{8x} - \frac{16x^2}{8x} - \frac{8x}{8x} = 3x^2 - 2x - 1$$

NOTE: $-8x$ divided by $8x = -1$ because any number (except zero) divided by itself is 1, and since the two numbers have opposite signs, the quotient is negative.

# 5.11 Parentheses

You have learned that parentheses are used to include terms, to group them and treat them as one whole number. Our next job is to simplify algebraic expressions involving parentheses.

*Examples:* **1.** $3(4x + 2) - 4(2x - 5)$

STEP 1 Use the distributive law.
$12x + 6 - 8x + 20$

STEP 2 Combine like terms.
$4x + 26$

This is $3(4x + 2) - 4(2x - 5)$ in simplified form.

**2.** $a^2 - (a^2 - 6)$

STEP 1 The minus sign before the parentheses means that $(a^2 - 6)$ is multiplied by $-1$.
$a^2 - a^2 + 6$

STEP 2 Combine like terms. We are left with:
6.

This is $a^2 - (a^2 - 6)$ in simplified form.

3. $(3x - 4)2 - (2x + 3)5$

STEP 1 Move the single-term factors 2 and 5 to the left of the binomial factors.
$2(3x - 4) - 5(2x + 3)$

STEP 2 Apply the distributive law.
$6x - 8 - 10x - 15$

STEP 3 Combine like terms.
$-4x - 23$

This is $(3x - 4)2 - (2x + 3)5$ in simplified form.

# 5.12 Evaluating Algebraic Expressions

To evaluate an expression means to replace the letters with numbers and simplify.

*Examples:* Evaluate the following expressions if $a = 3$, $b = -2$, and $c = 0$.

1. $a + 2b$
$= 3 + 2(-2)$
$= 3 + (-4)$
$= -1$

2. $a^2 - 3b$
$= 3^2 - 3(-2)$
$= 9 + 6$
$= 15$

3. $c[2a^2 + 3b^2]$
$= 0[2 \cdot 3^2 + 3(-2)^2]$
$= 0[18 + 12]$
$= 0$

Note that in the third example it was not necessary to find the sum of 18 and 12 inside the brackets, since zero times *any* number results in zero for the product

Exercises

1. $(2x + 3y) + (7x - 3y)$
2. $$\begin{array}{r} 2x + \phantom{3}y - 4z \\ + \; 4x - 3y + 7z \\ \hline \end{array}$$
3. $2(a - 3c)$
4. $3x(x^2 - 2x - 3)$
5. $(x - 2)(x - 3)$
6. $(3x - 4)(2x + 7)$

7. $\dfrac{8x^7}{2x^4}$

8. $\dfrac{4a^2 - 8a - 12}{2}$

9. $\dfrac{12a^4 - 8a^3 - 4a^2}{2a^2}$

Evaluate the following:

**10.** $x^2 - 2y$ if $x = 5$ and $y = 3$

**11.** $a^2 + ab$ if $a = 2$ and $b = -2$

**12.** $(x - y)(x + y)$ if $x = 13$ and $y = -13$

**13.** Find the value of $V$ in $V = lwh$ if $l = 8$, $h = 3$, and $w = 5$.

**14.** Find the value of $i$ in $i = prt$ if $p = 200$, $r = 0.06$, and $t = 4$.

**15.** Find the value of $a$ in $a = p + prt$ if $p = 200$, $r = 0.06$, and $t = 4$.

**Answers**

| | | |
|---|---|---|
| **1.** $9x$ | **6.** $6x^2 + 13x - 28$ | **11.** 0 |
| **2.** $6x - 2y + 3z$ | **7.** $4x^3$ | **12.** 0 |
| **3.** $2a - 6c$ | **8.** $2a^2 - 4a - 6$ | **13.** 120 |
| **4.** $3x^3 - 6x^2 - 9x$ | **9.** $6a^2 - 4a - 2$ | **14.** 48 |
| **5.** $x^2 - 5x + 6$ | **10.** 19 | **15.** 248 |

# Trial Test — Beginning Algebra

Before you begin the trial test, here are a few hints on how to take a multiple choice test.

1. Read the question carefully to understand what information is given.
2. Decide what process to use.
3. If no answer is given which agrees with yours, find your mistake.
4. Sometimes the answers will provide you with clues. Suppose a question is given as follows:
   Solve for $x$: $5x - 12 = 2x$
   (1) 2  (4) −4
   (2) 4  (5) 0
   (3) −2
   You could solve the equation and find the value of $x$ to be 4. (It would be a good idea to check this.) On the other hand, you could try each of the choices until you found one that worked.

**1.** Solve for $x$: $3x - 2 = -17$

(1) 5
(2) −5
(3) $-\dfrac{19}{3}$
(4) $\dfrac{19}{3}$
(5) 45

**1.** 1 2 3 4 5

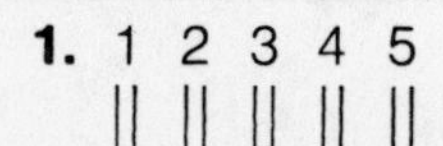

**2.** Solve for $y$: $\frac{y}{3} + 4 = -5$

(1) $-27$
(2) $-9$
(3) $27$
(4) $-3$
(5) $-1$

2. 1 2 3 4 5

**3.** $(-2)^3 =$

(1) 6
(2) $-6$
(3) $-8$
(4) 8
(5) 12

3. 1 2 3 4 5

**4.** Find the value of $x$ which satisfies the equation: $5x - 7 = 2x - 34$.

(1) $\frac{41}{3}$
(2) $-\frac{41}{3}$
(3) $3\frac{3}{7}$
(4) 9
(5) $-9$

4. 1 2 3 4 5

**5.** Find the product of $(a - 5)$ and $(a + 3)$.

(1) $a^2 - 15$
(2) $a^2 + 2a - 15$
(3) $a^2 - 2a - 15$
(4) $a^2 - 2$
(5) $2a - 2$

5. 1 2 3 4 5

**6.** What must $2x^3$ be multiplied by to make a product of $-6x^7$?

(1) $-12x^{10}$
(2) $-3x^4$
(3) $3x^4$
(4) $12x^{21}$
(5) $-12x^{21}$

6. 1 2 3 4 5

**7.** Divide $(8a^4 + 12a^3 - 24a^2)$ by $4a^2$.

(1) $2a^2 + 3a - 6$
(2) $2a^2 + 12a^3 - 24a^2$
(3) 10
(4) $2a^4 + 3a^3 - 24a^2$
(5) $2a^2 + 12a$

7. 1 2 3 4 5

**8.** Find the value of $2a^2 - 3b$, if $a = 2$ and $b = -3$.

(1) $-1$
(2) 1
(3) $-17$
(4) 17
(5) 12

8. 1 2 3 4 5

**9.** Evaluate $(a - b)(a + b)$, if $a = 20$ and $b = -3$.

(1) 37
(2) $-391$
(3) $-37$
(4) 532
(5) 391

9. 1 2 3 4 5

**10.** Solve for $z$: $3z - 5 + 2z = 25 - 5z$

(1) 0
(2) 3
(3) $-3$
(4) 1
(5) no solution

10. 1 2 3 4 5

**11.** $(-3)^4 + (-2)^4 + (-1)^4 =$

(1) 98
(2) $-98$
(3) $-21$
(4) 21
(5) $(-6)^4$

11. 1 2 3 4 5

**12.** Perform the indicated operations: $(-8) + (-3) - (7) - (-11) =$
(1) 15
(2) $-15$
(3) 7
(4) $-7$
(5) 3

**12.** 1 2 3 4 5

## SOLUTIONS

**1. (2)** Add 2 to each side to eliminate $-2$ from the left side. (You are "undoing" the subtraction.)

$$3x - 2 + 2 = -17 + 2$$
$$3x = -15$$

Now divide each side by 3 to find $x$. (You are "undoing" the multiplication.)

$$\frac{3x}{3} = \frac{-15}{3}$$
$$x = -5$$

**2. (1)** Subtract 4 from both sides to eliminate 4 from the left side. (This is the same as adding $-4$.)

$$\frac{y}{3} + 4 - 4 = -5 - 4$$
$$\frac{y}{3} = -9$$

Now multiply each side by 3 to find $y$.

$$3\left(\frac{y}{3}\right) = 3(-9)$$
$$y = -27$$

**3. (3)** $(-2)^3 = (-2)(-2)(-2)$, not $3(-2)$. Hence $(-2)^3 = -8$.

**4. (5)** Subtract $2x$ from each side to eliminate $2x$ from the right side.

$$5x - 7 - 2x = 2x - 34 - 2x$$
$$3x - 7 = -34$$

Now add 7 to each side to eliminate 7 from the left side.

$$3x - 7 + 7 = -34 + 7$$
$$3x = -27$$

Divide each side by 3 to find $x$.

$$\frac{3x}{3} = \frac{-27}{3}$$
$$x = -9$$

**5. (3)** $(a - 5)(a + 3)$
$= a \cdot a + 3 \cdot a + (-5) \cdot a + (-5) \cdot 3$

Simplify the products.

$$a^2 + 3a - 5a - 15$$

Now combine like terms.

$$a^2 - 2a - 15$$

**6. (2)** To find what must multiply $2x^3$ to form $-6x^7$, use the *inverse* operation — division:

$$\frac{-6x^7}{2x^3} = -3x^4$$

**7. (1)** $\dfrac{8a^4 + 12a^3 - 24a^2}{4a^2}$

$$= \frac{8a^4}{4a^2} + \frac{12a^3}{4a^2} - \frac{24a^2}{4a^2}$$
$$= 2a^2 + 3a - 6$$

**8. (4)** $2a^2 - 3b = 2(2^2) - 3(-3)$ when $a = 2$ and $b = -3$.

$$= (2)(4) + 9$$
$$= 8 + 9$$
$$= 17$$

**9. (5)** $(a - b)(a + b) = [20 - (-3)][20 + (-3)]$ when $a = 20$ and $b = -3$.

$$= (23)(17)$$
$$= 391$$

**10. (2)** $3z - 5 + 2z = 25 - 5z$

Combine like terms:

$$5z - 5 = 25 - 5z$$

Now add $5z$ to each side to eliminate $5z$ from the right side.

$$5z - 5 + 5z = 25 - 5z + 5z$$
$$10z - 5 = 25$$

Add 5 to each side to "undo" the subtraction.

$$10z - 5 + 5 = 25 + 5$$
$$10z = 30$$
$$z = 3$$

**11.** **(1)** $(-3)^4 + (-2)^4 + (-1)^4 = 81 + 16 + 1$
$= 98$

Remember that $(-3)^4 = (-3)(-3)(-3)(-3)$, not $4(-3)$.

**12.** **(4)** $(-8) + (-3) - (7) - (-11)$
$= -8 - 3 - 7 + 11$
$= -18 + 11$
$= -7$

## EVALUATION CHART

Multiply the number of correct answers by 100 and divide by the number of questions in the Trial Test to arrive at your score. ________

| Score | Rating |
|---|---|
| 90–100 | Excellent |
| 80–89 | Very Good |
| 70–79 | Average |
| 60–69 | Passing |
| 59 or Below | Very Poor |

# Chapter 6

# FACTORING IN ALGEBRA

# 6.1 Factoring

In previous work, you learned that the word *factor* means *multiplier*. Let us say that you are asked to *factor* the number 50. This means that you are to find numbers, which, when multiplied together, result in the product 50.

You could say: 50 = 1 times 50
*or* 50 = 2 times 25
*or* 50 = 5 times 10
*or* 50 = 2 times 5 times 5

In general, when we say *factor,* we mean that you should find only *prime* factors. A *prime number* is any number *except 1* which is divisible only by *itself* and *1*. So in the case of the different factorizations of 50 given above:

50 = (1)(50) is incorrect because neither 1 nor 50 is prime.

50 = (2)(25) is incorrect because 25 is not prime.

50 = (5)(10) is incorrect because 10 is not prime.

50 = (2)(5)(5) is the *correct* factorization because 2 and 5 are indeed prime numbers, each being divisible only by itself and 1, in accordance with our definition of a prime number.

*Examples:* Factor each of these expressions.

1. $12 = (2)(2)(3)$
2. $a^2b = a \cdot a \cdot b$
3. $10x^2y^3 = 2 \cdot 5 \cdot x \cdot x \cdot y \cdot y \cdot y$
4. $39a^3x^4 = 3 \cdot 13 \cdot a \cdot a \cdot a \cdot x \cdot x \cdot x \cdot x$

# 6.2 Finding the Common Factor

Factoring is a special kind of division, where we are given the answer to a multiplication problem, and we must find the factors which, when multiplied together, will result in the given product.

First, remember the distributive law: $a(b + c) = ab + ac$. Notice that the two terms in the product on the right-hand side each contain the factor $a$. So, if we are given $ab + ac$, and are told to factor this binomial, it's easy to see that $a$ is one factor, and of course, to find the other factor of the product we must divide each term by the factor we know, namely: $a$.

$$\frac{ab + ac}{a} = b + c. \quad \text{So: } ab + ac \text{ in factored form is } a(b + c).$$

*Examples:* **1.** Factor $75x^2 + 30x$.

STEP 1 Factor each term into its prime factors, so that you can find *all* the *common* factors and multiply them to find the *greatest common factor* (GCF).

$$75x^2 + 30x = 3 \cdot 5 \cdot 5 \cdot x \cdot x + 2 \cdot 3 \cdot 5 \cdot x$$

Common factors are 3, 5, and $x$.

$$3 \cdot 5 \cdot x = 15x$$

$15x$ is the GCF.

STEP 2 Divide each term by this GCF, $15x$, to find the other polynomial factor.

$$\frac{75x^2 + 30x}{15x} = 5x + 2 \text{ (the other factor)}$$

STEP 3 Write the given expression as the product of the GCF and the polynomial factor.

$$75x^2 + 30x = 15x(5x + 2)$$

**2.** Factor $36a^4 - 18a^3 + 24a^2$.

STEP 1 Factor each term into its prime factors.

$$\underline{2} \cdot 2 \cdot \underline{3} \cdot 3 \cdot \underline{a} \cdot \underline{a} \cdot a \cdot a - \underline{2} \cdot \underline{3} \cdot 3 \cdot \underline{a} \cdot \underline{a} \cdot a + \underline{2} \cdot 2 \cdot 2 \cdot \underline{3} \cdot \underline{a} \cdot \underline{a}$$

Common factors are 2, 3, $a$, $a$.

$$2 \cdot 3 \cdot a \cdot a = 6a^2$$

$6a^2$ is the GCF.

STEP 2 $(36a^4 - 18a^3 + 24a^2) \div 6a^2 = 6a^2 - 3a + 4$

STEP 3 $36a^4 - 18a^3 + 24a^2 = 6a^2(6a^2 - 3a + 4)$

It is important to practice many examples of factoring in order to understand clearly the idea and to develop some speed.

**Exercises** Factor the following:

1. $5x + 5$
2. $3y - 12$
3. $4xy + 4xz$
4. $a^2 - a$
5. $12 - 36y$
6. $x^4 + x^3 + 42x$
7. $b^3y - b^2 + b$
8. $\pi r^2 + 2\pi r$
9. $5x^2 - 10x + 20$
10. $abx^2 - 2acx + 5ax$

**Answers**

1. $5(x + 1)$
2. $3(y - 4)$
3. $4x(y + z)$
4. $a(a - 1)$
5. $12(1 - 3y)$
6. $x(x^3 + x^2 + 42)$
7. $b(b^2y - b + 1)$
8. $\pi r(r + 2)$
9. $5(x^2 - 2x + 4)$
10. $ax(bx - 2c + 5)$

# 6.3 Factoring the Difference of Two Squares

Take a look, now, at the products of some special pairs of binomials.

*Examples:*

1. $(a + b) \cdot (a - b) = (a + b)a - (a + b)b$    distributive law
$$= a^2 + ab - ab - b^2$$
$$= a^2 - b^2$$

2. $(x - 2y) \cdot (x + 2y) = (x - 2y)x + (x - 2y)2y$
$$= x^2 - 2xy + 2xy - 4y^2$$
$$= x^2 - 4y^2$$

3. $(x + 3) \cdot (x - 3) = (x + 3)x + (x + 3)(-3)$
$$= x^2 + 3x - 3x - 9$$
$$= x^2 - 9$$

The binomial factors with which we started in each case were special ones. One binomial was the *sum* of two numbers, while the other was the *difference* of the same two numbers. The product, in each case, was the *difference* between the *squares* of the *same* two numbers.

> The *special product* of the sum of two numbers and the *difference* of the *same two numbers* equals the *difference* of the *squares* of these numbers.
>
> $$(a + b)(a - b) = a^2 - b^2$$

**Exercises**

Use the rule above to find the following special products of the *sum* and *difference* of two numbers.

1. $(c + d)(c - d)$
2. $(k - m)(k + m)$
3. $(r + \pi)(r - \pi)$
   NOTE: $\pi$ (pi, pronounced "pie") is a literal number used in geometry.
4. $(z + 1)(z - 1)$
5. $(2a + 3b)(2a - 3b)$
6. $(5c^4 - 8d)(5c^4 + 8d)$
7. $(xy + 10)(xy - 10)$
8. $(6abc - 7)(6abc + 7)$
9. $(39)(41)$
   HINT: $39 = (40 - 1)$ and $41 = (40 + 1)$
10. $(48)(52)$

**Answers**

1. $c^2 - d^2$
2. $k^2 - m^2$
3. $r^2 - \pi^2$
4. $z^2 - 1$
5. $4a^2 - 9b^2$
6. $25c^8 - 64d^2$
7. $x^2y^2 - 100$
8. $36a^2b^2c^2 - 49$
9. $(40 - 1)(40 + 1) = 1{,}600 - 1 = 1{,}599$
10. $2{,}500 - 4 = 2{,}496$

Now that you see that the special product $(a - b)(a + b)$ is equal to $a^2 - b^2$, we can turn our attention to "undoing" the multiplication or *factoring* the expression $a^2 - b^2$. Recall that finding factors is the *inverse* or opposite process from multiplication.

To **factor** the *difference of two square numbers,* take the square root of the first number and the square root of the other number, and separate them, *once with a plus sign* and *once with a minus sign.*

$$a^2 - b^2 = (a + b)(a - b)$$

**Exercises** Use the rule above to factor the following binomials:

1. $x^2 - 36$
2. $64 - a^2$
3. $\pi^2 r^2 - 1$
4. $-x^2 + 1$
5. $\frac{x^2}{4} - \frac{a^2}{9}$
6. $0.04 - x^2$
7. $100 + k^2$
8. $6 - x^2$

**Answers**

1. $(x + 6)(x - 6)$
2. $(8 + a)(8 - a)$
3. $(\pi r + 1)(\pi r - 1)$
4. $-x^2 + 1 = 1 - x^2 = (1 + x)(1 - x)$
5. $\left(\frac{x}{2} - \frac{a}{3}\right)\left(\frac{x}{2} + \frac{a}{3}\right)$
6. $(0.2 + x)(0.2 - x)$
7. $100 + k^2$ cannot be factored by our rule, because it is *not* the *difference* between two squares.
8. $6 - x^2$, too, cannot be factored by our rule, because 6 is not a square number, and so $6 - x^2$ is not the difference between *two squares*.

# 6.4 Factoring Trinomials of the Form: $ax^2 + bx + c$

Examine, now, some products of two binomials which result in a trinomial product.

*Examples:*

1. $(x + 1)(x + 3) = x \cdot x + x(+3) + x(+1) + (+1)(+3)$
$= x^2 + 3x + x + 3$
$= x^2 + 4x + 3$

2. $(x + 1)(x - 3) = x \cdot x + x(-3) + x(+1) + (1)(-3)$
$= x^2 - 2x - 3$

3. $(x - 1)(x + 3) = x \cdot x + x(+3) + x(-1) + (-1)(+3)$
$= x^2 + 2x - 3$

4. $(x - 1)(x - 3) = x \cdot x + x(-3) + x(-1) + (-1)(-3)$
$= x^2 - 4x + 3$

Notice that, in *each* binomial, there is a *first* term and a *second* term. Our application of the distributive law shows us that, in each case:

1. We multiply the *first terms* in the two binomials to obtain the *first term* in our product. Call this first term $F$ (for *first*).

2. Next, we multiply the *outer terms*: The *first term* in the *first* binomial and the *last term* in the *last* binomial. Call this term $O$ (for *outer*).

3. For the third term we multiply the inner terms: the *last term* in the *first* binomial and the *first term* in the *last* binomial. Call this term $I$ (for *inner*). Since the $O$ and $I$ terms are both multiples of $x$, they can be combined to form a single term, $O + I$.

4. The fourth and last term in our product is found by multiplying the *last terms* in *the two binomials*. Call this term $L$ (for *last*).

5. The sequence *first, outer, inner,* and *last* gives us a short word, "*foil*," which will remind us how to multiply two binomials quickly.

Exercises — Multiply the following binomials, finding the individual terms as well as the trinomial product. (A few examples have been worked out.)

| BINOMIALS | F | O | I | L | TRINOMIAL PRODUCT |
|---|---|---|---|---|---|
| **1.** $(x + 1)(x + 5)$ | $x^2$ | $+5x$ | $+1x$ | $+5$ | $x^2 + 6x + 5$ |
| **2.** $(x + 2)(x + 5)$ | | | | | |
| **3.** $(x + 4)(x + 7)$ | | | | | |
| **4.** $(x + 8)(x + 3)$ | | | | | |
| **5.** $(x - 1)(x - 5)$ | $x^2$ | $-5x$ | $-1x$ | $+5$ | $x^2 - 6x + 5$ |
| **6.** $(x - 2)(x - 5)$ | | | | | |
| **7.** $(x - 4)(x - 7)$ | | | | | |
| **8.** $(x - 8)(x - 3)$ | | | | | |
| **9.** $(x + 1)(x - 5)$ | $x^2$ | $-5x$ | $+1x$ | $-5$ | $x^2 - 4x - 5$ |
| **10.** $(x + 2)(x - 5)$ | | | | | |
| **11.** $(x + 4)(x - 7)$ | | | | | |
| **12.** $(x + 8)(x - 3)$ | | | | | |
| **13.** $(x - 1)(x + 5)$ | $x^2$ | $+5x$ | $-1x$ | $-5$ | $x^2 + 4x - 5$ |
| **14.** $(x - 2)(x + 5)$ | | | | | |
| **15.** $(x - 4)(x + 7)$ | | | | | |
| **16.** $(x - 8)(x + 3)$ | | | | | |
| **17.** $(a + y)(2a + 3y)$ | $2a^2$ | $+3ay$ | $+2ay$ | $+3y^2$ | $2a^2 + 5ay + 3y^2$ |
| **18.** $(a - y)(2a - 3y)$ | | | | | |
| **19.** $(a + y)(2a - 3y)$ | | | | | |
| **20.** $(a - y)(2a + 3y)$ | | | | | |

**Answers**

1. $x^2 + 6x + 5$
2. $x^2 + 7x + 10$
3. $x^2 + 11x + 28$
4. $x^2 + 11x + 24$
5. $x^2 - 6x + 5$
6. $x^2 - 7x + 10$
7. $x^2 - 11x + 28$
8. $x^2 - 11x + 24$
9. $x^2 - 4x - 5$
10. $x^2 - 3x - 10$
11. $x^2 - 3x - 28$
12. $x^2 + 5x - 24$
13. $x^2 + 4x - 5$
14. $x^2 + 3x - 10$
15. $x^2 + 3x - 28$
16. $x^2 - 5x - 24$
17. $2a^2 + 5ay + 3y^2$
18. $2a^2 - 5ay + 3y^2$
19. $2a^2 - ay - 3y^2$
20. $2a^2 + ay - 3y^2$

Now let us look at some trinomials that are the products of *two binomials,* to see if we can come up with a rule to help us find the two unknown binomial factors.

Let us start with the four trinomials we obtained as products in the beginning of this section.

The trinomials we will look at are a special kind. They are called *quadratic trinomials.* You can identify them by their three terms using only one variable, whose highest power is 2.

| | F | O+I | L | | | F | O+I | L |
|---|---|---|---|---|---|---|---|---|
| **1.** | $x^2 +$ | $4x$ | $+ 3$ | | **3.** | $x^2 -$ | $2x$ | $- 3$ |
| **2.** | $x^2 -$ | $4x$ | $+ 3$ | | **4.** | $x^2 +$ | $2x$ | $- 3$ |

STEP 1 In each case, you remember that $F$ — the *first* term — came from the product of the *two* first terms in the binomials. So, by factoring $F$, we find the *first* terms to place in each set of parentheses. In each case here, $F$ is $x^2$, so write $x$ (the factor of $x^2$) in each set of parentheses.

**1.** $x^2 + 4x + 3 = (x \quad)(x \quad)$ **3.** $x^2 - 2x - 3 = (x \quad)(x \quad)$

**2.** $x^2 - 4x + 3 = (x \quad)(x \quad)$ **4.** $x^2 + 2x - 3 = (x \quad)(x \quad)$

STEP 2 Next we need to decide, if we can, what sign to put in each binomial factor after our $x$. Remember that $L$, the *last* term in our trinomial, was the *product* of the two *last* terms we are now trying to find to put into our binomials. Remember, too, that like signs (two minus signs or two plus signs) will result in a positive $L$. In other words, if the sign of $L$ is (+), it says that the signs of the two missing terms will be *alike*.

Also, because the middle term $(O + I)$ provides the *sum* of the last two terms, the signs of the two last terms will be not only *alike* but also *the same as the sign of the middle term.* (See examples 1 and 2 below.)

If, however, the sign of $L$ is $(-)$, we know that the signs in our missing binomials will be *unlike*. This says that one sign must be $(+)$ and the *other* must be $(-)$ (see examples 3 and 4).

**1.** (positive $L$, positive $O+I$)
$x^2 + 4x + 3 = (x + \quad)(x + \quad)$

**2.** (positive $L$, negative $O+I$)
$x^2 - 4x + 3 = (x - \quad)(x - \quad)$

**3.** (negative $L$, negative $O+I$)
$x^2 - 2x - 3 = (x + \quad)(x - \quad)$

**4.** (negative $L$, positive $O+I$)
$x^2 + 2x - 3 = (x + \quad)(x - \quad)$

In examples 1 and 2, the plus sign of the last term determines the signs in the two binomials to be the same. They will be either both $(+)$ or both $(-)$. In examples 3 and 4, the minus sign in $L$, which is $-3$, says our two signs must be unlike, which means one must be $(+)$ and one must be $(-)$.

STEP 3 Now all we have left to do is to find all sets of possible factors of $L$. Since $L$ in each case above is the number 3, and luckily 3 is a prime number, we have only two possibilities: 1 and 3.

Example 1 becomes:
$x^2 + 4x + 3 = (x + 1)(x + 3)$

Example 2 becomes:
$x^2 - 4x + 3 = (x - 1)(x - 3)$

In examples 3 and 4, we have a more difficult problem, since $L$ in these two problems is $-3$. There are two possibilities for sets of factors: $(+1)(-3)$ and $(-1)(+3)$.

Example 3: Try $-1$ and 3.
$x^2 - 2x - 3 = (x + 3)(x - 1)$

Example 4: Try $-1$ and 3.
$x^2 + 2x - 3 = (x + 3)(x - 1)$

In both cases, $O+I = +3x - 1x = +2x$, which is *correct* for example 4 but incorrect for example 3. So we must use the other possibility for example 3:

$$x^2 - 2x - 3 = (x - 3)(x + 1)$$

CHECK Multiply the binomial factors to make sure that their product is the trinomial given to be factored. You may use either the long multiplication or our short "*FOIL*" method. *But be sure to check. This step is most important.*

**1.** $(x + 1)(x + 3) = x^2 + 3x + 1x + 3 = x^2 + 4x + 3$

**2.** $(x - 1)(x - 3) = x^2 - 3x - 1x + 3 = x^2 - 4x + 3$

**3.** $(x + 1)(x - 3) = x^2 - 3x + 1x - 3 = x^2 - 2x - 3$

**4.** $(x + 3)(x - 1) = x^2 - 1x + 3x - 3 = x^2 + 2x - 3$

Below you will find a condensation of the above four steps, which you will find helpful in trying to find the two binomial factors of a trinomial of the form: $ax^2 + bx + c$.

TO FACTOR A TRINOMIAL:

1. Write two factors of the trinomial's *first term* as first terms in the two unknown binomials.
2. Look at the sign of the *last term* in the trinomial:
   If it is (+), write *like* signs in your binomials and make them *like* the sign of the *middle* term.
   If it is (−), write (+) in one binomial and (−) in the other.
3. Factor the *last* term of the trinomial into *all* possible pairs of factors and use the pair that will work to make $O+I$ equal the middle term of the trinomial.
4. Check by multiplication.

Look at the following illustrations, which will help you review the four steps given to factor trinomials:

*Examples:* **1.** Factor $x^2 - 3x - 10$.

STEP 1 $(x \quad )(x \quad )$

STEP 2 Last term $(-10)$ is negative, so:

$$(x + \quad )(x - \quad )$$

STEP 3 Possible pairs of factors of $-10$ are: $(+1,-10)$, $(-1,+10)$, $(+2,-5)$, $(-2,+5)$. Try $(x + 2)(x - 5)$.

CHECK
$$(x + 2)(x - 5) = \overset{F}{x^2} - \overset{O}{5x} + \overset{I}{2x} - \overset{L}{10}$$
$$= x^2 - 3x - 10$$

**2.** Factor $x^2 - 11x + 30$.

STEP 1 $(x \quad )(x \quad )$

STEP 2 Last term is $+30$ (so both signs are *alike*) and *middle term* is $-11x$ (so both missing numbers are *negative*):

$$(x - \quad )(x - \quad )$$

STEP 3 Possible pairs of factors of $+30$ are: $(-1,-30)$, $(-2,-15)$, $(-3,-10)$, $(-5,-6)$. (We omit the positive factors, since we now know that both missing numbers are negative.) Try $(x - 5)(x - 6)$.

CHECK
$$(x - 5)(x - 6) = \overset{F}{x^2} - \overset{O}{6x} - \overset{I}{5x} + \overset{L}{30}$$
$$= x^2 - 11x + 30$$

The only complication comes when the first term is not $x^2$, but $ax^2$, where $a$ is some number other than $+1$. This type of problem is handled the same way, but usually requires many more trials. To repeat, there is no substitute for experience, which comes from practice on *many* problems.

*Example:* Factor $2x^2 + 7x - 15$.

STEP 1 $(x \quad )(2x \quad )$

STEP 2 Last term is $-15$, so either: $(x + \quad )(2x - \quad )$
or: $(x - \quad )(2x + \quad )$

STEP 3 Possible pairs of factors of $-15$ are: $(+1,-15)$, $(+3,-5)$, $(+5,-3)$, $(+15,-1)$, $(-1,+15)$, $(-3,+5)$, $(-5,+3)$, $(-5,+3)$.

STEP 4 Using the first pair of factors, $(+1,-15)$, we have $(x + 1)(2x - 15)$, which gives us a middle term of $-15x + 2x$, or $-13x$. This is wrong; the middle term must be $+7x$. Using the third pair of factors, $(x + 5)(2x - 3)$, we get the middle term $-3x + 10x$, or $+7x$. Thus $(+5,-3)$ is the right choice.

CHECK
$$(x + 5)(2x - 3) = \overset{F}{2x^2} - \overset{O}{3x} + \overset{I}{10x} - \overset{L}{15} = 2x^2 + 7x - 15$$

**Exercises**

1. $x^2 + 3x + 2$
2. $x^2 - 3x + 2$
3. $x^2 + x - 2$
4. $x^2 - x - 2$
5. $x^2 + 8x + 12$
6. $x^2 - 8x + 12$
7. $x^2 + 4x - 12$
8. $x^2 - 4x - 12$
9. $x^2 + 7x + 12$
10. $x^2 - 7x + 12$
11. $x^2 + x - 12$
12. $x^2 - x - 12$
13. $x^2 + 13x + 12$
14. $x^2 + 11x - 12$
15. $x^2 - 14x + 40$
16. $x^2 - 6x - 40$
17. $x^2 + 22x + 40$
18. $x^2 + 18x - 40$
19. $x^2 + 39x - 40$
20. $x^2 - 2x - 80$
21. $x^2 - 2x - 63$
22. $x^2 - 2x - 24$
23. $x^2 - 2x - 48$
24. $x^2 - x - 42$
25. $2x^2 + 19x + 35$
26. $4x^2 - 11x + 6$
27. $3x^2 + 17x + 10$
28. $2x^2 + x - 6$
29. $2x^2 - x - 6$
30. $2x^2 - 5x - 7$

**Answers**

1. $(x + 1)(x + 2)$
2. $(x - 1)(x - 2)$
3. $(x + 2)(x - 1)$
4. $(x + 1)(x - 2)$
5. $(x + 2)(x + 6)$
6. $(x - 2)(x - 6)$
7. $(x + 6)(x - 2)$
8. $(x + 2)(x - 6)$
9. $(x + 3)(x + 4)$
10. $(x - 3)(x - 4)$
11. $(x + 4)(x - 3)$
12. $(x + 3)(x - 4)$
13. $(x + 1)(x + 12)$
14. $(x + 12)(x - 1)$

**15.** $(x-4)(x-10)$ **23.** $(x+6)(x-8)$
**16.** $(x+4)(x-10)$ **24.** $(x+6)(x-7)$
**17.** $(x+2)(x+20)$ **25.** $(x+7)(2x+5)$
**18.** $(x+20)(x-2)$ **26.** $(x-2)(4x-3)$
**19.** $(x+40)(x-1)$ **27.** $(x+5)(3x+2)$
**20.** $(x+8)(x-10)$ **28.** $(x+2)(2x-3)$
**21.** $(x+7)(x-9)$ **29.** $(x-2)(2x+3)$
**22.** $(x+4)(x-6)$ **30.** $(x+1)(2x-7)$

# 6.5 Complete Factoring

When we say *factor* now, we mean *factor completely*, using all three methods when, and if, they apply. Let us list these again:

REVIEW OF FACTORING METHODS:

1. Find the *greatest common factor* (GCF) in the expression and then divide by this GCF to obtain the polynomial factor.

$$\frac{ab+ac}{a}=b+c$$

2. Recognize the *difference between two squares* and break the given binomial into two other binomials, one of which is the sum of the *square roots* of the given squares, and the other the *difference* of these *same square* roots.

$$(a^2-b^2)=(a+b)(a-b)$$

3. Break certain quadratic trinomials into two binomial factors.

| Trinomial | Binomial factors |
|---|---|
| (+) last term, (+) middle term: | $(x+\quad)(x+\quad)$ |
| (+) last term, (−) middle term: | $(x-\quad)(x-\quad)$ |
| (−) last term, (−) middle term: | $(x+\quad)(x-\quad)$ |
| (−) last term, (+) middle term: | $(x+\quad)(x-\quad)$ |

**How to Factor Completely:**

STEP 1 *First see if there is a monomial factor.*

**a.** If so, remove it before trying either of the other two methods.

**b.** If not, go on to step 2.

*Examples:* **1.** Factor $a^6+10a^4-56a^2$ (GCF is $a^2$)
$=a^2(a^4+10a^2-56)$. Now go to step 2.

**2.** Factor $x^4+8x^2-9$.

In this second illustration we have no common factor, so we go at once to step 2.

STEP 2 *Now count the terms in the polynomial factor.*

**a.** If it has but *two terms*, you can factor it only if it is the *difference of two squares.*

**b.** If it has *three terms* as in each of the quadratic trinomials above, try to *find two binomials that will give you the desired product.* From the first example above we have:

$a^2(a^4 + 10a^2 - 56)$
$= a^2(a^2 + 14)(a^2 - 4)$
$= a^2(a^2 + 14)(a + 2)(a - 2)$

Note that the binomial $(a^2 - 4)$ was *not* factored completely, since it was the difference of two squares. In our second illustration above, we proceed at once to step 2, and we have:

$x^4 + 8x^2 - 9$
$= (x^2 + 9)(x^2 - 1)$
$= (x^2 + 9)(x + 1)(x - 1)$

STEP 3 *Write all the factors, and, of course, be sure to check.*

*Examples:* (The steps in parentheses refer to the preceding instructions on factoring.)

**1.** Factor $6x^2 + 15x + 6$.
GCF is 3.
$3(2x^2 + 5x + 2)$ (STEP 1)
$3(x + 2)(2x + 1)$ (STEP 2)
CHECK $3(x + 2)(2x + 1)$ (STEP 3)
$3(2x^2 + 1x + 4x + 2)$
$3(2x^2 + 5x + 2)$
$6x^2 + 15x + 6$

**2.** Factor $4a^2 - 36$.
GCF is 4.
$4(a^2 - 9)$ (STEP 1)
$4(a + 3)(a - 3)$ (STEP 2)
CHECK $4(a + 3)(a - 3)$ (STEP 3)
$4(a^2 - 3a + 3a - 9)$
$4(a^2 - 9)$
$4a^2 - 36$

**3.** Factor $2x^4 + 162$.
GCF is 2.
$2(x^4 + 81)$ (STEP 1)
NOTE: $x^4 + 81$ is *not* the difference of two squares, and so you *cannot* factor it. (STEP 2)
CHECK $2(x^4 + 81)$ (STEP 3)
$2x^4 + 162$

**4.** Factor $2x^2 + 2x + 4$.
GCF is 2.
$2(x^2 + x + 2)$ (STEP 1)
$2(x + 1)(x + 2)$ (STEP 2)

Here we have the only possible binomials $(x + 1)(x + 2)$ and we see they produce $x^2 + 2x + x + 2$, which equals $x^2 + 3x + 2$, and this $\neq x^2 + x + 2$.

CHECK $2(x^2 + x + 2)$ must be our answer, since it produces the original polynomial: $2x^2 + 2x + 4$. (STEP 3)

**Exercises**

1. $3x^3 - 27x$
2. $3x^2 - 27x$
3. $2w^2 + 36w + 162$
4. $8z^2 + 112z + 392$
5. $6x^2 - 11x + 5$
6. $64x^2 - 16x + 1$
7. $25y^4 - 144z^2$
8. $2x^2 + 10x - 168$
9. $2x^2 - 2$
10. $s^2 + 4s - 45$
11. $k^2 + 15k - 16$
12. $a^2x^2 + 9ax + 14$
13. $x^4 + x^2 - 20$
14. $x^6 - 7x^4 + 12x^2$
15. $2x^2 + 10x + 10$
16. $\pi R^2 - \pi r^2$
17. $16a^4 - b^4$
18. $u^2 + 35u - 36$
19. $2y^2 + 7y + 3$
20. $25x^2 + 90xy + 81y^2$

**Answers**

1. $3x(x + 3)(x - 3)$
2. $3x(x - 9)$
3. $2(w + 9)(w + 9)$ or $2(w + 9)^2$
4. $8(z + 7)(z + 7)$ or $8(z + 7)^2$
5. $(x - 1)(6x - 5)$
6. $(8x - 1)(8x - 1)$
7. $(5y^2 + 12z)(5y^2 - 12z)$
8. $2(x + 12)(x - 7)$
9. $2(x + 1)(x - 1)$
10. $(s + 9)(s - 5)$
11. $(k + 16)(k - 1)$
12. $(ax + 2)(ax + 7)$
13. $(x^2 + 5)(x + 2)(x - 2)$
14. $x^2(x^2 - 3)(x + 2)(x - 2)$
15. $2(x^2 + 5x + 5)$
16. $\pi(R - r)(R + r)$ Note here that $R$ and $r$ represent two different numbers, so we cannot say that either one is a common factor of the expression; $\pi$ is the only common factor.
17. $(4a^2 + b^2)(2a + b)(2a - b)$
18. $(u + 36)(u - 1)$
19. $(y + 3)(2y + 1)$
20. $(5x + 9y)(5x + 9y)$

# Trial Test — Factoring

In the following examples, factor each expression into *prime* factors.

**1.** $x^2 - 4$
(1) $(x - 2)(x - 2)$
(2) $(x - 2)(x + 2)$
(3) $(x + 2)(x + 2)$
(4) $(x - 4)(x + 1)$
(5) $(x + 4)(x - 1)$

**1.** 1 2 3 4 5

**2.** $5a^2 - 125$
(1) $(5a - 25)(a + 5)$
(2) $5(a^2 - 25)$
(3) $(5a + 25)(a - 5)$
(4) $5(a - 5)(a + 5)$
(5) $(a + 5)(a + 5)$

**2.** 1 2 3 4 5

**3.** $y^2 - y - 6$
(1) $(y - 3)(y + 2)$
(2) $(y + 3)(y - 2)$
(3) $(y + 2)(y + 3)$
(4) $(y - 3)(y - 2)$
(5) $(y - 6)(y + 1)$

**3.** 1 2 3 4 5

**4.** $m^2 + 7m + 12$
(1) $(m + 6)(m + 2)$
(2) $(m + 10)(m + 2)$
(3) $(m + 6)(m + 1)$
(4) $(m + 1)(m + 12)$
(5) $(m + 4)(m + 3)$

**4.** 1 2 3 4 5

**5.** $6a^2 - 12a + 36$
(1) $6(a - 2)(a + 3)$
(2) $(6a - 3)(a - 12)$
(3) $6(a - 3)(a + 2)$
(4) $6(a^2 - 2a + 6)$
(5) $(3a - 6)(2a - 6)$

**5.** 1 2 3 4 5

**6.** $x^4 - 2x^3 - 15x^2$
(1) $x^2(x - 5)(x + 3)$
(2) $(x^3 - 5x^2)(x + 3)$
(3) $(x^3 + 3x^2)(x - 5)$
(4) $x^2(x^2 - 2x - 15)$
(5) $(x^2 - 5x)(x^2 + 3x)$

**6.** 1 2 3 4 5

**7.** $5x^3 - 5x$
(1) $5x(x - 1)(x + 1)$
(2) $5x(x^2 - 1)$
(3) $5(x^3 - x)$
(4) $(5x + 5)(x^2 - 1)$
(5) $(5x - 5)(x^2 + 1)$

**7.** 1 2 3 4 5

**8.** $2x^2 + x - 6$
(1) $(2x + 3)(x - 2)$
(2) $(x + 2)(x - 3)$
(3) $(x + 3)(x - 2)$
(4) $(2x + 1)(x - 6)$
(5) $(2x - 3)(x + 2)$

**8.** 1 2 3 4 5

**9.** $4m^2 - 16n^2$
(1) $(4m - 8n)(m + n)$
(2) $(2m + 4n)(2m - 4n)$
(3) $4(m - 2n)(m + 2n)$
(4) $2(m^2 + n^2)$
(5) $4(m^2 - 4n^2)$

**9.** 1 2 3 4 5

**10.** $2x^2 - 10xy - 28y^2$
(1) $2(x - 7y)(x + 2y)$
(2) $(2x + 4y)(x - 7y)$
(3) $(2x - 14y)(x + 2y)$
(4) $2(x - 7y)(x + 2y)$
(5) $2(x^2 - 5xy - 14y^2)$

**10.** 1 2 3 4 5

## SOLUTIONS

**1.** **(2)** $(x - 2)(x + 2) = x^2 - 4$

**2.** **(4)** $5(a - 5)(a + 5) = 5(a^2 - 25)$
$= 5a^2 - 125$

**3.** **(1)** $(y - 3)(y + 2) = y^2 - y - 6$

**4.** **(5)** $(m + 4)(m + 3) = m^2 + 7m + 12$

**5.** **(4)** $6(a^2 - 2a + 6) = 6a^2 - 12a + 36$
(NOTE: $a^2 - 2a + 6$ is prime.)

**6.** **(1)** $x^2(x - 5)(x + 3) = x^2(x^2 - 2x - 15)$
$= x^4 - 2x^3 - 15x^2$

**7.** **(1)** $5x(x - 1)(x + 1) = 5x(x^2 - 1)$
$= 5x^3 - 5x$

**8.** **(5)** $(2x - 3)(x + 2) = 2x^2 + x - 6$

**9.** **(3)** $4(m - 2n)(m + 2n) = 4(m^2 - 4n^2)$
$= 4m^2 - 16n^2$

**10.** **(1)** $2(x - 7y)(x + 2y) = 2(x^2 - 5xy - 14y^2)$
$= 2x^2 - 10xy - 28y^2$

## EVALUATION CHART

Multiply the number of correct answers by 100 and divide by the number of questions in the Trial Test to arrive at your score. ________

| Score | Rating |
|---|---|
| 90–100 | Excellent |
| 80–89 | Very Good |
| 70–79 | Average |
| 60–69 | Passing |
| 59 or Below | Very Poor |

# Chapter 7

# ALGEBRA — MORE ON EQUATIONS

All the work in the preceding chapters leads to the solution of problems. If algebra is to be of any value at all, it is as a tool to solve problems. To use this tool, any problem must first be *translated* into an *equation* that can be *solved*. Before you can solve a problem involving an algebraic solution, you must be able to solve some elementary equations.

> REMEMBER:
>
> Any equation can be solved by use of one or more of the four fundamental operations.

# 7.1 Simultaneous Equations

A slightly more complicated type of problem involves *two unknowns*. Suppose we have an equation such as:

$$x + y = 10$$

What is the solution? If we let $x = 7$, and $y = 3$, the statement is true. Also, if $x$ is 0 and $y$ is 10, the statement is true. Then, too, another solution would be: $x = -1$, and $y = 11$. You can see that there seems to be no end to the number of solutions.

There is no unique solution to an equation containing two variables. But, by introducing a *second* equation, you can search for a solution that satisfies both equations at the same time (simultaneously). These two equations are known as simultaneous equations.

*Examples:* **1.** $x + y = 10$
$x - y = 4$

STEP 1 It is necessary to find two numbers, one for $x$, and one for $y$, so that both statements are true *at the same time*. The trick here is to get rid of one of the variables. This can be done, in this particular pair, by *adding the two equations together*:

$$\begin{array}{rl} x + y &= 10 \\ x - y &= \ \ 4 \\ \hline 2x \qquad &= 14 \\ x \qquad &= \ \ 7 \end{array}$$

(The sum of $+y$ and $-y$ is 0.)
(Dividing by 2.)

STEP 2 Since the addition eliminates $y$, the resulting equation has only one variable and is easily solved. To find the value of $y$, simply replace $x$ by 7 in either equation.

$$\begin{array}{rl} x + y &= 10 \\ 7 + y &= 10 \\ y &= \ \ 3 \end{array} \quad \text{or} \quad \begin{array}{rl} x - y &= 4 \\ 7 - y &= 4 \\ 7 &= 4 + y \\ 3 &= y \\ y &= 3 \end{array}$$

So the complete solution is: $x = 7$
$y = 3$

These two numbers make *each equation true.*

CHECK

$$\begin{array}{rl} x + y &= 10 \\ 7 + 3 &\stackrel{?}{=} 10 \\ 10 &= 10 \end{array} \quad \textit{and} \quad \begin{array}{rl} x - y &= 4 \\ 7 - 3 &\stackrel{?}{=} 4 \\ 4 &= 4 \end{array}$$

**2.** $5x + y = 17$
$3x + y = 11$

STEP 1 This time, adding the two equations won't eliminate either variable, but subtracting them will eliminate $y$.

*Adding:*

$$\begin{array}{rl} 5x + \ \ y &= 17 \\ 3x + \ \ y &= 11 \\ \hline 8x + 2y &= 28 \end{array}$$

*Subtracting:*

$$\begin{array}{rl} 5x + y &= 17 \\ 3x + y &= 11 \\ \hline 2x + 0 &= \ \ 6 \\ x &= \ \ 3 \end{array}$$

STEP 2 Substitute 3 for $x$ in one of the equations.

$$\begin{array}{rl} 5(3) + y &= 17 \\ 15 + y &= 17 \\ y &= \ \ 2 \end{array}$$

So the solution is: $\left.\begin{array}{l} x = 3 \\ y = 2 \end{array}\right\}$

**3.** $3x + 2y = 19$
$2x - \ \ y = \ \ 8$

STEP 1 This time, neither adding nor subtracting will eliminate either variable. One or both of the equations must be changed so that one variable can be removed. In the first equation we have $3x$, while in the second we have $2x$. Both can be converted to $6x$ if the first equation is multiplied by 2 and the second equation is multiplied by 3. We usually show the multipliers of each equation in the following way:

$$2/\ 3x + 2y = 19 \rightarrow 6x + 4y = 38$$
$$3/\ 2x - \ \ y = \ \ 8 \rightarrow 6x - 3y = 24$$

STEP 2 Now subtract the second equation from the first equation to eliminate $x$.

$$\begin{array}{r} 6x + 4y = 38 \\ -\ (6x - 3y = 24) \\ \hline 7y = 14 \\ y = \ \ 2 \end{array}$$

STEP 3 Substitute 2 for $y$:

$$\begin{aligned} 3x + 2(2) &= 19 \\ 3x + 4 &= 19 \\ 3x &= 15 \\ x &= \ \ 5 \end{aligned}$$

So the complete solution is: $\left.\begin{array}{l} x = 5 \\ y = 2 \end{array}\right\}$

Exercises

1. $x + y = 12$
   $x - y = \ \ 6$
2. $3a + b = 7$
   $a + b = 1$
3. $2x + 3y = 1$
   $x + \ \ y = 1$
4. $3c - 2d = 13$
   $2c + 3d = 0$

Answers

1. $\left.\begin{array}{l} x = 9 \\ y = 3 \end{array}\right\}$
2. $\left.\begin{array}{l} a = 3 \\ b = -2 \end{array}\right\}$
3. $\left.\begin{array}{l} x = 2 \\ y = -1 \end{array}\right\}$
4. $\left.\begin{array}{l} c = 3 \\ d = -2 \end{array}\right\}$

# 7.2 Quadratic Equations

A *quadratic equation* is an equation where the highest power of the variable is 2. Here are some examples:

$$x^2 + x - 6 = 0$$
$$3x^2 = 5x - 7$$
$$x^2 - 4 = 0$$
$$64 = x^2$$

How can we solve equations like these?

First, look at this little equation: $A \cdot B = 0$. If $A$ and $B$ stand for numbers, what can we say about them? Surely one of them must be zero, or else their product could not possibly be zero as stated in $A \cdot B = 0$.

REMEMBER:

If $A \cdot B = 0$, then either $A = 0$ *or* $B = 0$.

You might ask what this has to do with quadratic equations. To answer, look at this equation:

$$x^2 + x - 6 = 0$$

Thinking back to factoring, it's possible to write $x^2 + x - 6$ as $(x + 3)(x - 2)$.

If $\quad x^2 + x - 6 = 0$
then $\quad (x + 3)(x - 2) = 0$.

Now we have two factors, and either one or both may be zero.

$x + 3 = 0$ and/or $x - 2 = 0$
$x = -3$ and/or $x = 2$

We must check these possible solutions in the *original* equation.

When $x = -3$: $(-3)^2 + (-3) - 6 \stackrel{?}{=} 0$
$+9 - 3 - 6 \stackrel{?}{=} 0$
$0 = 0$

When $x = 2$: $(2)^2 + (2) - 6 \stackrel{?}{=} 0$
$4 + 2 - 6 \stackrel{?}{=} 0$
$0 = 0$

TO SOLVE A QUADRATIC EQUATION:

1. Place all terms on one side of the equation, so that all that remains on the other side is zero.
2. Factor the polynomial.
3. Set each factor equal to zero.
4. Solve the resulting equations.
5. Check your answers in the *original* equation.

*Example:* Solve $x^2 = 3x + 10$.

STEP 1 Place all terms on one side.

$$x^2 - 3x - 10 = 0$$

STEP 2 Factor the polynomial.

$$(x + 2)(x - 5) = 0$$

STEP 3 Set each factor equal to zero and solve the equations.

$x + 2 = 0 \qquad x - 5 = 0$
$x = -2 \qquad x = 5$

CHECK Substitute each answer in the original equation.

$x^2 = 3x + 10$
$(-2)^2 \stackrel{?}{=} 3(-2) + 10$
$4 \stackrel{?}{=} -6 + 10$
$4 = 4$

$x^2 = 3x + 10$
$(5)^2 \stackrel{?}{=} 3(5) + 10$
$25 \stackrel{?}{=} 15 + 10$
$25 = 25$

**Exercises** Here are some quadratic equations to solve:

1. $x^2 + 4x + 3 = 0$
2. $x^2 = 16$
3. $x^2 - 2x - 48 = 0$
4. $x^2 - 25 = 0$
5. $2a^2 + a - 6 = 0$
6. $y^2 = y + 72$
7. $x^2 - 14 = 5x$
8. $x^2 - 7x + 12 = 0$

**Answers**

1. $x = -1$, $x = -3$
2. $x = 4$, $x = -4$
3. $x = 8$, $x = -6$
4. $x = 5$, $x = -5$
5. $a = -2$, $a = \frac{3}{2}$
6. $y = 9$, $y = -8$
7. $x = -2$, $x = 7$
8. $x = 3$, $x = 4$

# 7.3 Inequalities

An inequality is a statement that two quantities are *not* equal to each other. When two quantities are not equal to each other, then one of two statements must be true of them:

1. Either the first must be *greater than* the second
2. or the first must be *less than* the second.

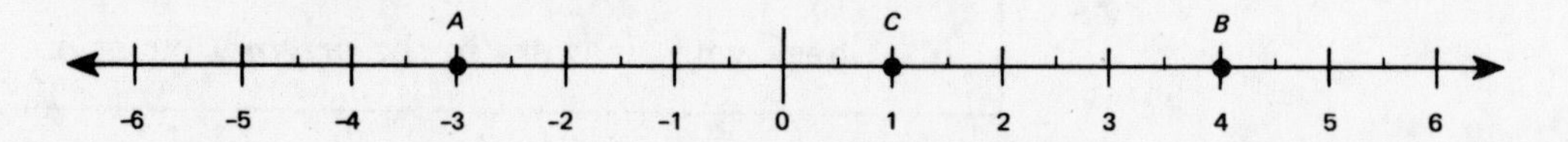


On the number line above, $A$ is to the *left* of $C$, and $C$ is to the *left* of $B$. Whenever $A$ is to the *left* of $B$, we say that *$A$ is less than $B$*. So you can see from our number line that $-3$ is less than 1, and 1 is less than 4; of course, it follows that $-3$ is less than 4.

We can make a few generalizations here:

1. Any negative number is less than zero.
2. Zero is less than any positive number.
3. Any negative number is less than any positive number.

Just as we have the symbol = to tell you that one number is equal to another, so do we have symbols of inequality.

| Symbols | Meanings |
|---|---|
| $6 \neq 7$ | 6 *does not equal* 7 |
| $7 \neq 6$ | 7 *does not equal* 6 |
| $7 > 6$ | 7 *is greater than* 6 |
| $7 < 8$ | 7 *is less than* 8 |
| $x \geq 8$ | $x$ *is greater than or equal to* 8 |
| $x \leq 8$ | $x$ *is less than or equal to* 8 |

**Exercises**

Think of a number line to help you determine whether the following statements are true or false:

1. $-2 < -3$
2. $2 \neq 1 + 1$
3. $-6 < 1$
4. $-0.5 < -0.75$
5. $0 > -15$
6. $0.5 > -2$
7. $-30 < -5$
8. $-30 < 5$
9. $-3 > 1$
10. $3 > 1$

**Answers**

1. False
2. False
3. True
4. False
5. True
6. True
7. True
8. True
9. False
10. True

Similar rules to those used for solving equations can be used to solve inequalities. Compare the procedures for solving equations and inequalities in the following illustrations:

*Examples:* 1.

EQUATION

$$\begin{aligned} x - 3 &= 8 \\ +3 &= +3 \\ \hline x &= 11 \end{aligned}$$

Add 3 to each side.

This means 11 is the *only* number which makes the equation $x - 3 = 8$ a *true* statement.

CHECK $x - 3 \stackrel{?}{=} 8$
$x = 11$

INEQUATION OR INEQUALITY

$$\begin{aligned} x - 3 &< 8 \\ +3 &= +3 \\ \hline x &< 11 \end{aligned}$$

Add 3 to each side.

This means that any number *less* than 11 when used for $x$ will make the statement $x - 3 < 8$ true. For example, let $x = 10$ and we have $10 - 3$, or 7, which is certainly less than 8.

*Examples:*

| | EQUATION | INEQUATION OR INEQUALITY |
|---|---|---|
| **2.** | $x + 5 = 7$ <br> $-5 = -5$ Add $-5$ to each side. <br> $x = 2$ <br> CHECK $2 + 5 \stackrel{?}{=} 7$ <br> $7 = 7$ | $x + 5 > 7$ <br> $-5 = -5$ Add $-5$ to each side. <br> $x > 2$ <br> Our statement $x + 5 > 7$ is true for any value of $x$ greater than 2. Let $x = 3$, and we have: $3 + 5$ or 8, which is greater than 7. |
| **3.** | $3x = 12$ <br> $\left(\frac{1}{3}\right)3x = 12\left(\frac{1}{3}\right)$ <br> Multiply both sides by $\frac{1}{3}$. <br> $x = 4$ <br> CHECK $3(4) \stackrel{?}{=} 12$ <br> $12 = 12$ | $3x < 12$ <br> $\left(\frac{1}{3}\right)3x < 12\left(\frac{1}{3}\right)$ <br> Multiply both sides by $\frac{1}{3}$. <br> $x < 4$ <br> Again: $3x < 12$ is true for *all* values of $x$ less than 4. Try 3: $3(3) = 9$ and $9 < 12$. |
| **4.** | $\frac{2}{3}x = 2$ <br> $\left(\frac{3}{2}\right)\frac{2}{3}x = 2\left(\frac{3}{2}\right)$ <br> Multiply both sides by $\frac{3}{2}$. <br> $x = 3$ <br> CHECK $\frac{2}{3}(3) \stackrel{?}{=} 2$ <br> $2 = 2$ | $\frac{2}{3}x \geqslant 2$ <br> $\left(\frac{3}{2}\right)\frac{2}{3}x \geqslant 2\left(\frac{3}{2}\right)$ <br> Multiply both sides by $\frac{3}{2}$. <br> $x \geqslant 3$ <br> Once more, $\frac{2}{3}x \geqslant 2$ is true for any value of $x$ that is greater than *or* equal to 3. Try: $x = 3$ and $x = 6$. <br> $\frac{2}{3}(3) \geqslant 2$  $\frac{2}{3}(6) \geqslant 2$ <br> $2 = 2$  $4 \geqslant 2$ |

There is one situation in solving inequalities which has no corresponding situation in equations.

Consider this statement: $7 \ ? \ 4$

Multiply each side by $-2$: $(-2)7 \ ? \ 4(-2)$

$-14 \ ? \ -8$

Is it true that $-14$ is *greater than* $-8$? If you will picture these as they appear on a number line, you will see that $-14$ is to the left of $-8$. Therefore, we find that, by multiplying by a negative number, we have *reversed* the original relationship. This means that while 7 is *greater* than 4, $-14$ turns out to be *less* than $-8$.

RULES TO REMEMBER FOR SOLVING INEQUALITIES:

1. The same number may be added to or subtracted from each side of an inequality.
2. Each side of an inequality may be multiplied or divided by the same *positive* number.
3. Each side of an inequality may be multiplied or divided by the same negative number, if, and *only* if, you remember to *reverse the inequality*.

*Examples:* **1.** Solve for $x$:

$$3x - 2 < 13$$

STEP 1 Add 2 to each side.

$$3x - 2 + 2 < 13 + 2$$
$$3x < 15$$

STEP 2 Divide both sides by 3 to find $x$.

$$\frac{3x}{3} < \frac{15}{3}$$
$$x < 5$$

**2.** Solve for $y$:

$$3 - 2y \geqslant 11$$

STEP 1 Subtract 3 from each side.

$$3 - 2y - 3 \geqslant 11 - 3$$
$$-2y \geqslant 8$$

STEP 2 Divide each side by $-2$.

$$\frac{-2y}{-2} \geqslant \frac{8}{-2}$$
$$y \leqslant -4$$

STEP 3 Reverse the inequality.

$$y \leqslant -4$$

Exercises

Find the answer or *solution set* for problems 1 through 4. (The *solution set* means the set of values that make the statement true.)

**1.** $3x - 5 < 7$

**2.** $2x + 3 \geqslant x + 8$

**3.** $-7x + 7 > -4x + 22$

**4.** $3x + 5 < x - 5$

Before trying the true or false problems (5–10), remember that, in mathematics, a statement is true if and only if the statement is *always* true. Otherwise the sentence is *false*. To say that a sentence is *false*, we need show *only one* case that will make it so.

For example, consider the statement:

$$\text{If } a \cdot b > 0, \text{ then } a > 0 \text{ and } b > 0.$$

Using $a = -1$ and $b = -2$, $a \cdot b > 0$. But $a < 0$ and $b < 0$. Thus the sentence is *false*, even though it works when positive numbers are used.

Answer true or false:

**5.** If $x \neq 0$, then $x^2 > 0$.

**6.** If $x > y$ and $y > z$, then $x > z$.

**7.** If $a > b$ and $b > 0$, then $\frac{1}{a} < \frac{1}{b}$.

**8.** $-3 > -2$

**9.** If $x^2 > 0$, then $x > 0$.

**10.** If $x - y > 0$, then $x > y$.

**Answers**

**1.** $x < 4$

**2.** $x \leq 5$

**3.** $x < -5$

**4.** $x < -5$

**5.** True, because if $x \neq 0$, then $x$ is either positive or negative, and in either case, $x^2$ will be positive, or greater than zero.

**6.** True.

**7.** True; try $a = 3$ and $b = 2$.

**8.** False, because $-3$ is at the left of $-2$ on the number line.

**9.** False, for if $x^2 = 4$, then $x$ can be $-2$, and $-2 < 0$.

**10.** True; add $y$ to each side.

# 7.4 Graphs of Equations

A <u>graph</u> of an equation is a set of points that make the equation true. You might think of it as a picture of an equation. For example, the graph of $x = \{-3,0,5\}$ is pictured on the horizontal number line below.

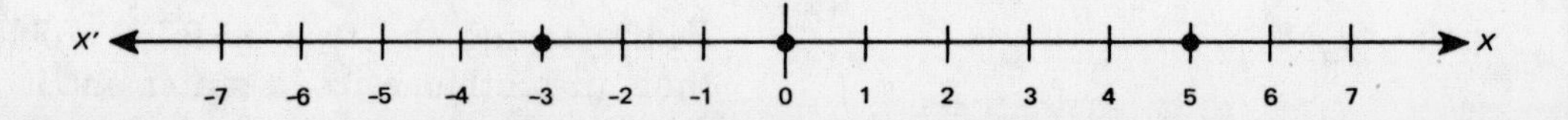

The graph of the equation $x = 3$ is shown below.

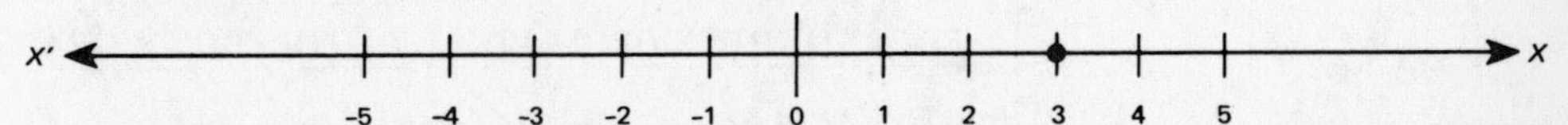


These two examples are illustrations of plotting points in one dimension. We need only one axis, or number line. To understand the plotting of points in *two* dimensions, we use two number lines, one horizontal and the second vertical.

These two *axes* (plural of *axis*) intersect at a point called the origin because it is our starting point, and represents the pair of numbers (0,0). The *first zero* tells us to move neither to the *right* nor to the *left* on the *horizontal* axis. The *second zero* tells us to move neither *up* nor *down* along the *vertical* axis.

By custom we call the first (*horizontal*) axis the X-axis and the second (*vertical*) axis the Y-axis. These two axes divide the graph paper into *four regions,* called quadrants.

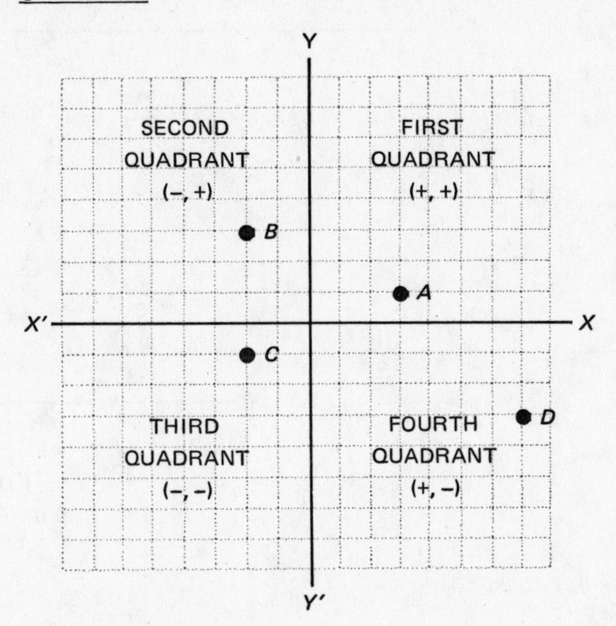


The *first quadrant* is the part *above the X-axis* and *to the right of the Y-axis.* Every point in it represents a *pair of positive numbers*, such as (+3,+1) (point *A* on the graph).

The *second quadrant* also lies *above the X-axis,* but *to the left of the Y-axis.* Every point in it represents a pair of numbers, the *first negative* and the *second positive;* for example, (−2,+3) (point *B* on the graph).

The *third quadrant* lies *to the left of the Y-axis*, but *below the X-axis*. Every point in it represents a *pair of negative numbers,* such as (−2,−1) (point *C* on the graph).

The *fourth quadrant,* like the first, lies *to the right of the Y-axis,* but unlike the first lies *below the X-axis*. Every point in it represents a pair of numbers, the *first a positive* number and the *second a negative* number. The number pair (+7,−3) (point *D* on the graph) is an example.

The pairs of numbers just described are called ordered pairs, because the order *does* make a difference. They are also called the coordinates of the point they represent.

FINDING THE COORDINATES OF A POINT ON A GRAPH:

*To find the first number (x):*

1. Count the number of spaces from the *Y-axis*, left or right, to reach the point. Remember that distance to the *right* is *positive* and distance to the *left* is *negative*.
2. If a point is *on the Y-axis*, its *x-value* or first number is *zero*.

*To find the second number (y):*

1. Count the number of spaces from the *X-axis*, up or down, to reach the point. Remember that distance *up* is *positive* and distance *down* is *negative*.
2. If a point is *on the X-axis*, its *y-value* or second number is *zero*.

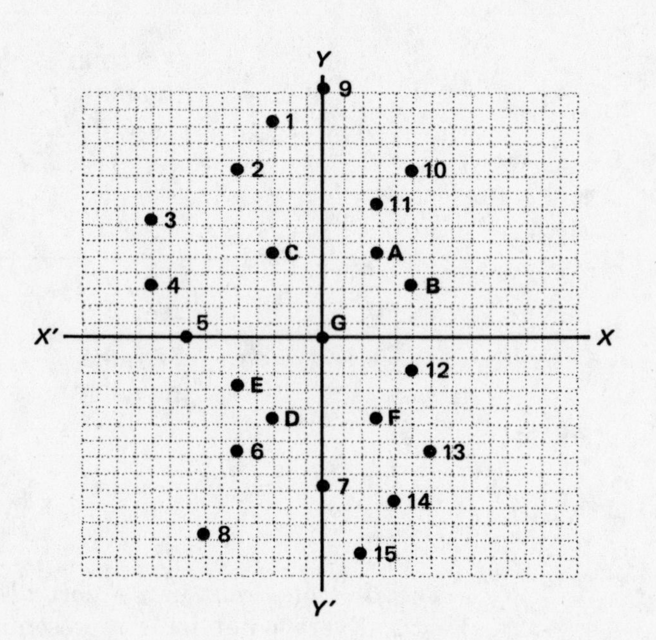


Now let us use our rules to find the coordinates of the points pictured on the preceding graph.

Point *A*: *To the right 3 spaces*, so the first number (*x-value*) is *+3*. *Up 5 spaces,* so the second number (*y-value*) is *+5*. *Pair:* (3,5)

Point *B*: *Right 5 spaces,* so the first number is *+5*. *Up 3 spaces,* so the second number is *+3*. *Pair:* (5,3)
NOTE: (3,5) and (5,3) represent *different* points.

Point *C*: *Left 3 spaces*, so the first number is *−3*. *Up 5 spaces,* so the second number is *+5*. *Pair:* (−3,5)

Point *D*: *Left 3 spaces,* so the first number is *−3*. *Down 5 spaces,* so the second number is *−5*. *Pair:* (−3,−5)

Point *E*: *Left 5 spaces,* so the first number is *−5*. *Down 3 spaces,* so the second number is *−3*. *Pair:* (−5,−3)
NOTE: (−3,−5) and (−5,−3) represent *different* points.

Point *F*: *Right 3 spaces*, so the first number is *3*. *Down 5 spaces,* so the second number is *−5*. *Pair:* (3,−5)

Point *G*: *Not left and not right*, so the first number is *0*. *Not up and not down*, so the second number is *0*. This is the *origin* and its coordinates are: (0,0).

**Exercises** Try to find the coordinates for each of the 15 numbered points plotted on the graph.

**Answers**

**1.** (−3,13) **2.** (−5,10) **3.** (−10,7) **4.** (−10,3) **5.** (−8,0)
**6.** (−5,−7) **7.** (0,−9) **8.** (−7,−12) **9.** (0,15) **10.** (5,10)
**11.** (3,8) **12.** (5,−2) **13.** (6,−7) **14.** (4,−10) **15.** (2,−13)

You remember that the equation $x = 3$ was represented by a point on one number line. But what is $x = 3$ in two-dimensional space with two number lines? In other words, what points have 3 for their $x$-value?

A look at the first graph below will tell you.

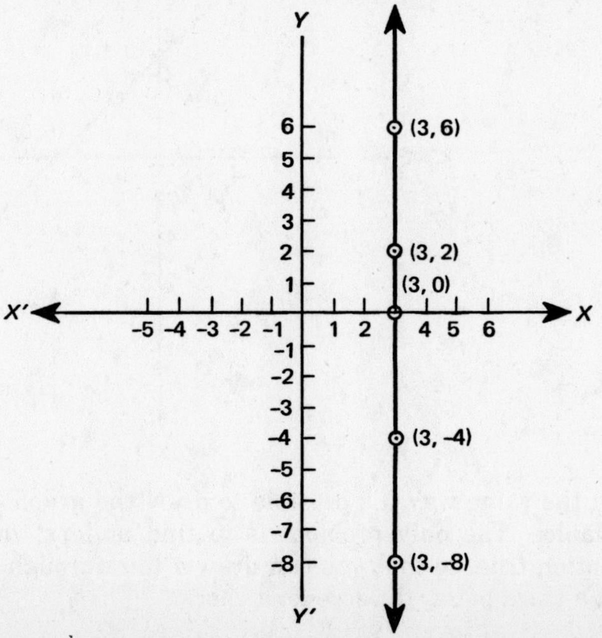


These points can be connected to form a *straight line*. This line represents all the points where the $x$-value is 3.
NOTE: There is a continuous line because an infinite number of points have an $x$-value of 3.
*For example:* (3, 4.5), (3, 4.58), (3, 4.58752)

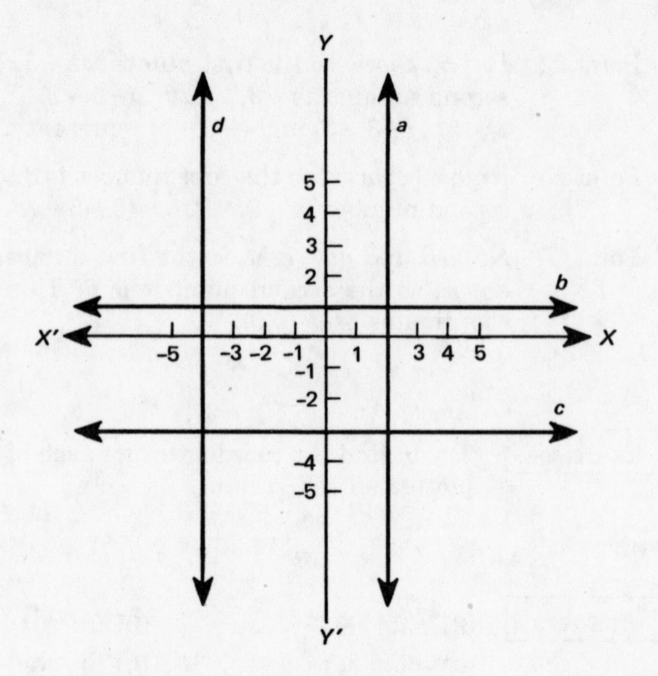


In the graph at left:
Line $a$ represents all the points where $x = 2$.
Line $b$ represents all the points where $y = 1$.
Line $c$ represents all the points where $y = -3$.
Line $d$ represents all the points where $x = -4$.
The $X$-axis represents all the points where $y = 0$.
The $Y$-axis represents all the points where $x = 0$.

Now, think about the equation $x + y = 5$. What pairs of numbers used for $x$ and $y$ will make the statement true? Some are: $(-5,10)$, $(0,5)$, $(2,3)$, $(5,0)$ and $(8,-3)$. On the graph below we have plotted these points and drawn the line connecting them. *The coordinates of any point on the line* will make the equation $x + y = 5$ true.

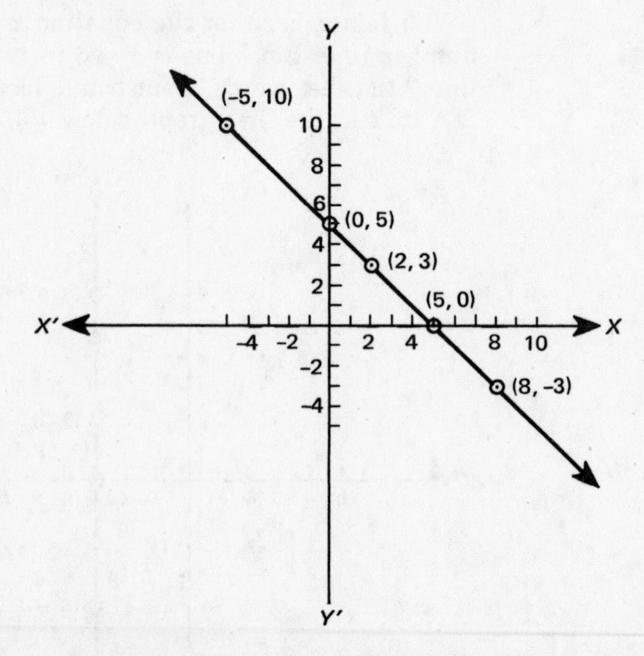


In the same way it's possible to draw the graph of any equation with two variables. The only problem is to find at least *two* points that make the equation true, so that you can draw a line through them. It is a good idea to plot a third point, to serve as a check.

*Example:* $2x + y = 7$

STEP 1 To make the problem easier, solve for $y$:

$$y = -2x + 7$$

STEP 2 Substitute *any* value for $x$ and obtain a corresponding value for $y$. Zero is usually an easy value to use for $x$. Find two more values.

*Substituting:*

If $x = 0, y = -2(0) + 7 = 7$
If $x = 1, y = -2(1) + 7 = 5$
If $x = 5, y = -2(5) + 7 = -3$

| $x$ | $y$ |
|---|---|
| 0 | 7 |
| 1 | 5 |
| 5 | −3 |

STEP 3 Plot the points and draw a line through them.

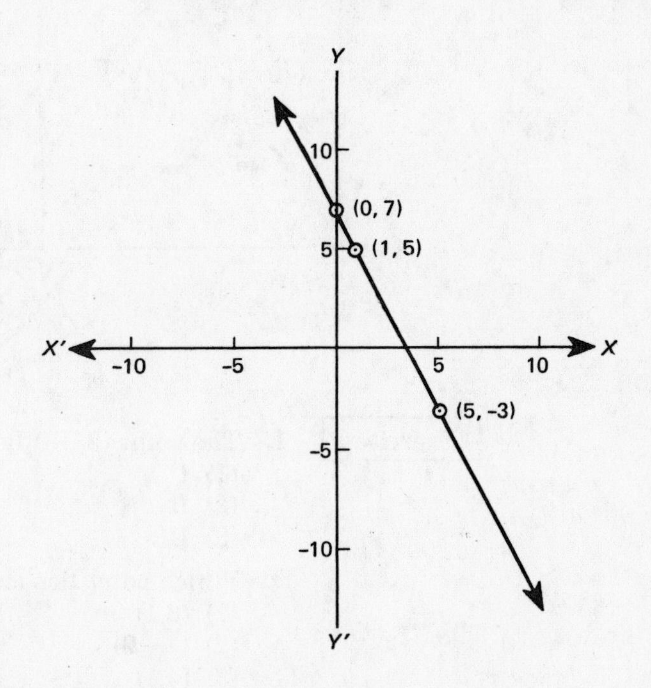


The equation $2x + y = 7$ is called a <u>*linear equation,*</u> since its graph is a straight line. It contains two variables, *not* multiplied, where each has an exponent of *one*.

<u>TO GRAPH A LINEAR EQUATION:</u>

1. Solve the equation for $y$.
2. Choose 3 values for $x$.
3. Find the 3 corresponding values for $y$.
4. Plot the points.
5. Join them to form a straight line.

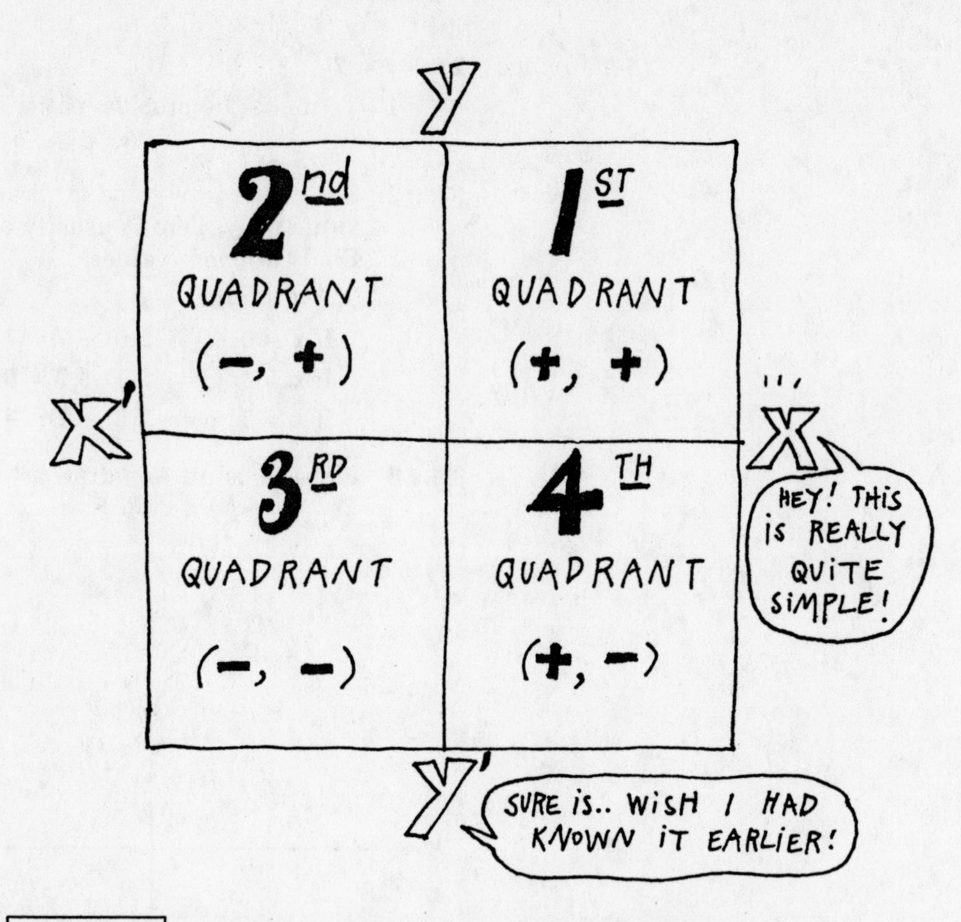


**Exercises**

1. The point (2,−3) lies in which quadrant?
   (1) I
   (2) II
   (3) III
   (4) IV
   (5) V
2. Which point lies on the line $x + 3y = 7$?
   (1) (3,1)
   (2) (1,−2)
   (3) (4,1)
   (4) (1,4)
   (5) (7,1)
3. The point (2,3) lies on which line?
   (1) $x + y = 7$
   (2) $x + y = 5$
   (3) $3x + 2y = 13$
   (4) $2x - 3y = 0$
   (5) $x + y = 3$
4. Which of the following is a horizontal line?
   (1) $x = 2$
   (2) $y = 2$
   (3) $x + y = 2$
   (4) $x = 0$
   (5) $2x - 2y = 0$
5. Which line passes through the origin?
   (1) $x + y = 1$
   (2) $2x - y = 3$
   (3) $x = -y$
   (4) $3x - 4y = 7$
   (5) $x - 2 = 0$

**Answers** 1. 4 2. 3 3. 2 4. 2 5. 3

# 7.5 Number Sequences

A popular type of question asked on many tests is the number sequence. The general form of the question is to give you some numbers, and ask which would come next.

There are many varieties of sequences, but in general, only common sense and a little reasoning are needed to come up with a correct answer.

Let's take a look at several examples, and the type of questions that might be asked.

*Examples:*

| | SEQUENCE | NEXT TERM | WHY |
|---|---|---|---|
| **1.** | 2,4,6,8, ? | 10 | Terms increase by 2. |
| **2.** | 1,4,7,10, ? | 13 | Terms increase by 3. |
| **3.** | 2,4,8,16, ? | 32 | Each term is twice the one before it. |
| **4.** | 12,6,3, ? | 1½ | Each term is one half the one before it. |
| **5.** | 1,4,3,6,5, ? | 8 | This is harder, but the pattern is up 3, down 1, up 3, down 1, etc. |
| **6.** | 1,1,2,3,5,8, ? | 13 | Again, a trick. Each term is the sum of the two terms preceding it. |
| **7.** | 11,7,4,2, ? | 1 | The difference between the first two terms is 4, between the second and the third is 3, between the third and the fourth is 2. Therefore, decrease by 1. |
| **8.** | 4,9,16,25, ? | 36 | These are the squares of 2, 3, 4, and 5. So the next term is $6^2$ or 36. |
| **9.** | 3,9,4,8, ? | 5 | This requires perception. The pattern is: up 6, down 5, up 4. Then the next term must be down 3. |

**Exercises** In each case determine the next term in the sequence:

**1.** 1,5,9,13, ?

**2.** 1,3,9,27, ?

**3.** 16,8,4,2, ?

**4.** 17,12,8,5, ?

**5.** 1,1,2,4,8, ?

**6.** 1,8,27, ?

**7.** 12,−6,3, ?

**8.** 2,7,3,6, ?

**9.** 2,6,18, ?

**10.** 4,7,11,16, ?

Answers

1. 17 2. 81 3. 1 4. 3 5. 16 6. 64 7. $-1\frac{1}{2}$ 8. 4 9. 54 10. 22

# Trial Test — More on Equations

**1.** If $a = 2$ and $b = -3$, then $3a + 2b = ?$
(1) 0
(2) $-5$
(3) 12
(4) $-12$
(5) 8

**1.** 1 2 3 4 5

**2.** If $x = -3$ and $y = -2$, then the value of $x^2y$ is
(1) $-8$
(2) 18
(3) $-18$
(4) $-11$
(5) 11

**2.** 1 2 3 4 5

**3.** The straight line $3x - 2y = 5$ will pass through which of the following points?
(1) (3,2)
(2) (2,3)
(3) (3,−2)
(4) (−3,2)
(5) (5,4)

**3.** 1 2 3 4 5

**4.** The solution set for $x^2 - x - 56 = 0$ is
(1) {7,8}
(2) {−7}
(3) {8}
(4) {−7,8}
(5) {7,−8}

**4.** 1 2 3 4 5

**5.** What positive integers are in the solution set for $3x - 1 < 7$?
(1) {0,1,2,3}
(2) {0,1,2}
(3) {0,1}
(4) {1,2}
(5) {1,2,3}

**5.** 1 2 3 4 5

**6.** If $x$ is a whole number, which value for $x$ will make $7 < 3x < 9$ a true statement?
(1) 8
(2) 2
(3) 0
(4) 1
(5) none of the above

**6.** 1 2 3 4 5

**7.** If $x$ is a negative number and $y$ is a negative number, then which of the following is a negative number?
(1) $xy$
(2) $x^2$
(3) $x^2y$
(4) $(-x)(-y)$
(5) $x^2y^2$

**7.** 1 2 3 4 5

**8.** If $x + 2y = 7$ and $x - 2y = 3$, then the value of $x$ is
(1) 2
(2) 5
(3) 4
(4) 3
(5) 1

**8.** 1 2 3 4 5

**9.** On the line $3x - 2y = 7$, the point with an $x$-value of 1 has a $y$-value of
(1) 2
(2) $-2$
(3) 5
(4) $-5$
(5) 0

9. 1 2 3 4 5

**10.** If $a < b$ and $c < d$, then
(1) $a < c$
(2) $a < d$
(3) $b < c$
(4) $b < d$
(5) There are insufficient data to reach a conclusion.

10. 1 2 3 4 5

## SOLUTIONS

**1. (1)** Use 2 for $a$, and $-3$ for $b$.
$$(3)(2) + 2(-3) = 6 + (-6) \text{ or } 0.$$

**2. (3)** $x^2y = x \cdot x \cdot y$

**3. (1)** Try 3 for $x$ and 2 for $y$.
$$3 \cdot 3 - 2 \cdot 2 = 5$$
$$9 - 4 = 5$$
$$5 = 5$$

**4. (4)** Factor $x^2 - x - 56$ and set each factor equal to zero.
$$x - 8 = 0 \qquad x + 7 = 0$$
$$x = 8 \qquad x = -7$$

**5. (4)** Since the word *positive* is used, 0 may not be a solution. This throws out the first three choices. When 3 is used for $x$, $3x - 1 = 8$, which is *not* less than 7. Therefore choice (4) is correct.

**6. (5)** Again, try all possible solutions. None will work, so the answer is choice (5).

**7. (3)** $x^2y$ means $x \cdot x \cdot y$. Since each is negative, the odd number of minus signs tells you that the product is negative.

**8. (2)** Add the two equations:
$$x + 2y = 7$$
$$x - 2y = 3$$
$$2x = 10$$
$$x = 5$$

**9. (2)** If $3x - 2y = 7$ and $x = 1$, then $3 - 2y = 7$. Solve this equation to find $y = -2$.

**10. (5)** There is no way to relate $a$ with $c$ or $d$, nor can you relate $b$ with $c$ or $d$.

## EVALUATION CHART

Multiply the number of correct answers by 100 and divide by the number of questions in the Trial Test to arrive at your score. ____________

| Score | Rating |
|---|---|
| 90–100 | Excellent |
| 80–89 | Very Good |
| 70–79 | Average |
| 60–69 | Passing |
| 59 or Below | Very Poor |

# Chapter 8

# GEOMETRY

Of all branches of mathematics, geometry is one of the most practical. It has to do with the world around us. Some knowledge of geometry is necessary for everyone. The word *geometry* means *earth measure*. In this section we will deal with practical problems involving measurement. Arithmetic and algebra will be used extensively in solving geometric problems.

As a first step, we must understand certain common words. Some words have special meanings in geometry. The words *point* and *line* you already understand, since you use them every day, but how about the word *angle*? Almost always a picture will help to describe the word.

# 8.1 Angles

Look at the diagram:

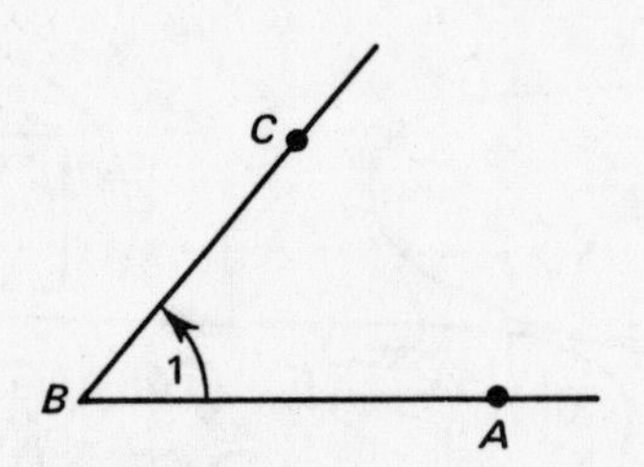


This is an <u>*angle*</u>. The symbol for the word *angle* is ∠. You can see that an angle is formed by two lines meeting at the point *B*. This point *B* is called the <u>*vertex*</u> of the angle. You can name the angle in three ways:

1. By the point at the vertex: ∠ *B*
2. By three letters, the vertex in the middle:
   ∠ *ABC* or ∠ *CBA* but *not* ∠ *BAC* or ∠ *CAB*
3. By a number inside the angle: ∠ 1.

To measure an angle, we use an instrument called a *protractor*.

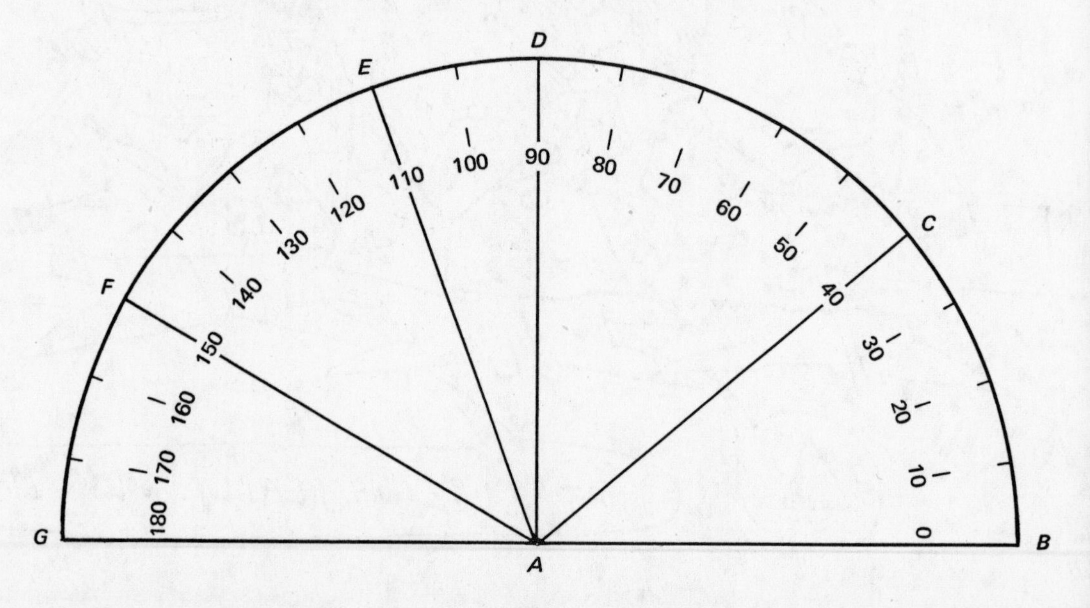


From the markings on the protractor you can find the size of the angle. Angles are measured in <u>*degrees*</u> (° is the degree symbol).

$\angle BAC = 40°$, $\angle BAE = 110°$, $\angle BAG = 180°$, $\angle BAD = 90°$, $\angle BAF = 150°$

How large an angle is $\angle CAD$? It starts at 40° and ends at 90°, so its size is 50°.

**Classification of Angles by Size**

**Acute angle:** an angle more than 0° and less than 90°.

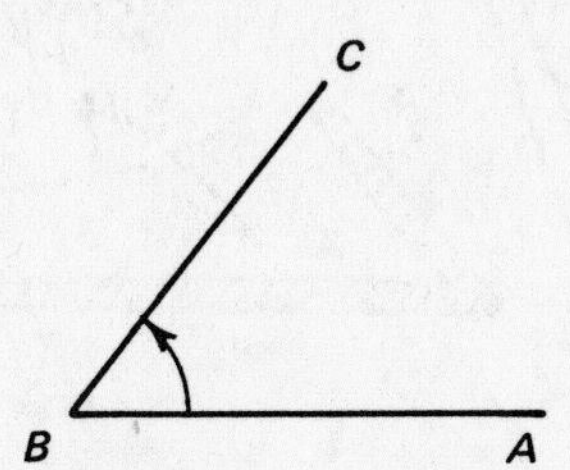


$\angle ABC$ is an *acute* angle.

**Right angle:** an angle of exactly 90°.

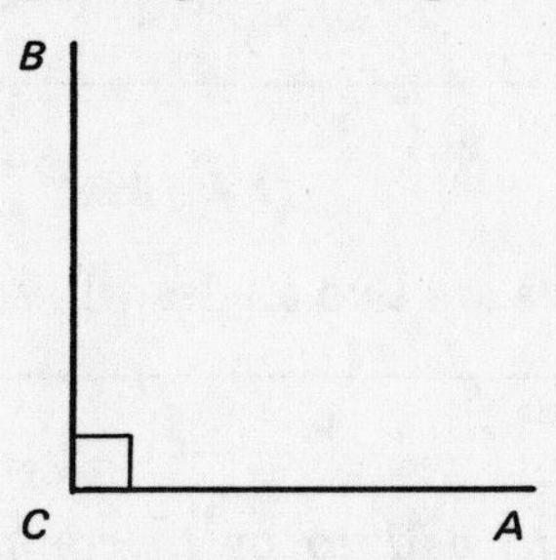


$\angle ACB$ is a *right* angle.

**Obtuse angle:** an angle more than 90° but less than 180°.

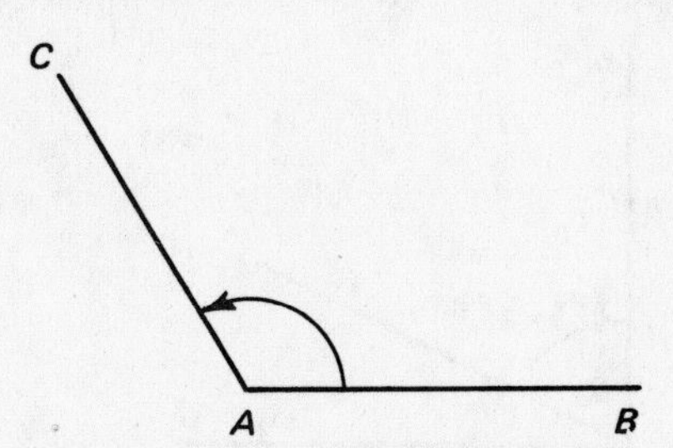


$\angle BAC$ is an *obtuse* angle.

**Straight angle:** an angle of exactly 180°.

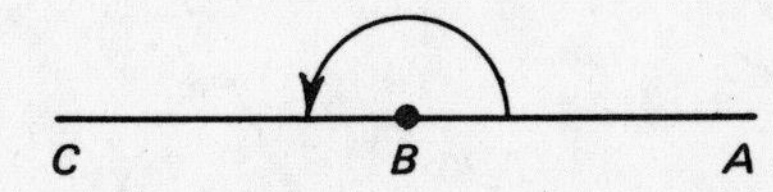


$\angle ABC$ is a *straight* angle.

The *degree* is the basic unit for measuring angles. Just as a yard is broken up into feet and inches, a degree is broken up into minutes (′) and seconds (″). Be careful not to confuse these minutes with *minutes of time.* The two ideas are similar in one respect, though; there are 60 *minutes* in one degree and 60 *minutes* in one hour.

One degree (1°) = 60 minutes (60′)
One minute (1′) = 60 seconds (60″)

*Pairs* of angles also have special names.

**DEFINITION**

**Complementary angles are two angles whose sum is 90°.**

*Complement* means *to add to* or *to complete.* (Don't confuse this with *compliment*, as in to pay a compliment to someone.)

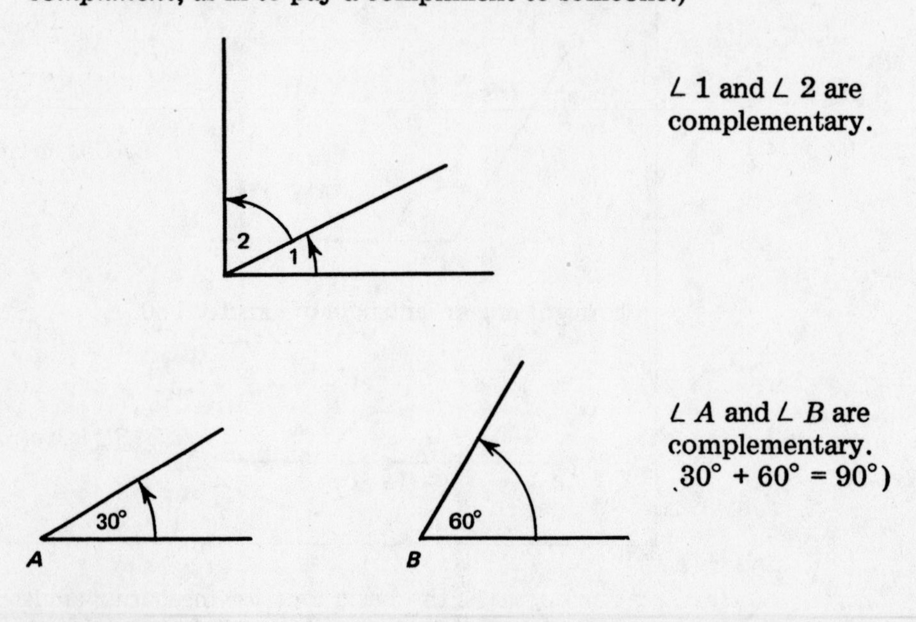


∠ 1 and ∠ 2 are complementary.

∠ *A* and ∠ *B* are complementary. (30° + 60° = 90°)

**DEFINITION**

**Supplementary angles are two angles whose sum is 180°.**

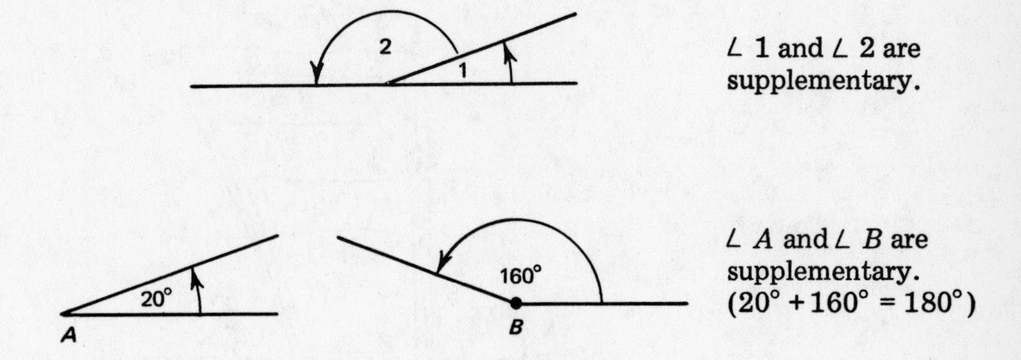


∠ 1 and ∠ 2 are supplementary.

∠ $A$ and ∠ $B$ are supplementary. ($20° + 160° = 180°$)

Let's see how to use algebra to solve a problem involving angles.

*Example*: Two angles are supplementary and one is 20° more than three times the other. Find the two angles.

STEP 1. Identify the unknowns with a letter.
Let:

$$x = \text{the smaller angle.}$$
$$3x + 20 = \text{the larger angle.}$$

(The $3x + 20$ comes directly from the problem: "20° more than three times the other angle.")

STEP 2. Use an additional fact from the problem to set up an equation. The additional fact in this case is that the two angles are *supplementary*, adding up to 180°.

$$x + 3x + 20 = 180°$$

STEP 3. Solve the equation.

$$x + 3x + 20° = 180°$$
$$4x + 20° = 180°$$
$$x = 40° \quad \text{(smaller angle)}$$
$$3(40°) + 20° = 140° \text{ (larger angle)}$$

# 8.2 Parallel and Perpendicular Lines

We now use angles to look at some special pairs of lines. These two special pairs of lines are *parallel* and *perpendicular* lines.

DEFINITION

Parallel lines are two lines which do not meet no matter how far extended.

$a$ $m$

$b$ $n$

$a \parallel b$ means line $a$ is *parallel* to line $b$.

Lines $m$ and $n$ are *not* parallel, because they would meet if they were extended.

> DEFINITION
>
> **Perpendicular lines are two lines which meet to form a right angle.**

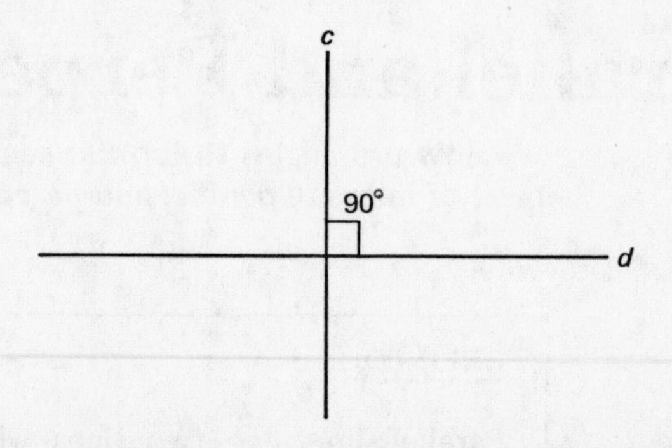


$c \perp d$ means line $c$ is perpendicular to line $d$. They meet at an angle of 90°.

# 8.3 Triangles

Most of you know that a triangle is a geometric figure with three straight lines. It is important to be able to classify triangles.

**Classification of Triangles by Their Sides**

**Equilateral triangle:** All three sides have the same length (and all three *angles* are equal).

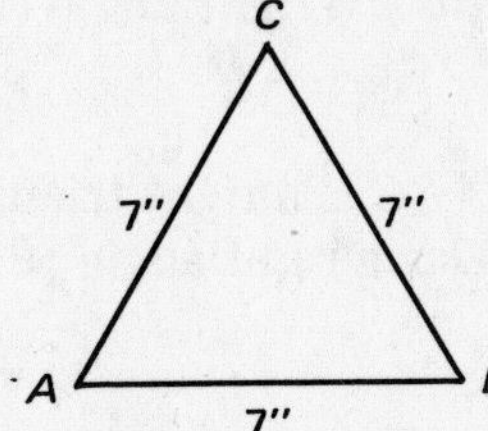


Triangle $ABC$ ($\triangle ABC$) is *equilateral.*
$AB = BC = AC$, and $\angle A = \angle B = \angle C$.

**Isosceles triangle:** Two sides have the same length (and the angles opposite the equal sides are equal).

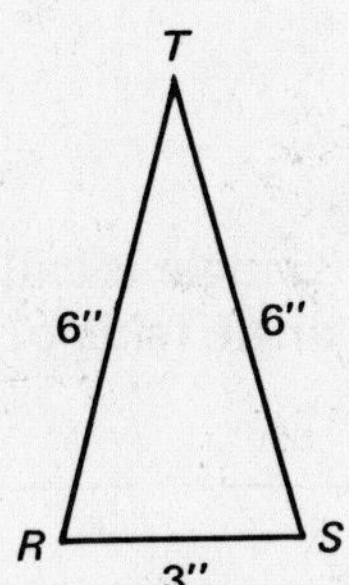


$\triangle RST$ is *isosceles.*
$RT = ST$, and $\angle R = \angle S$.

**Scalene triangle:** *No* two sides have the same length (and no two angles are equal).

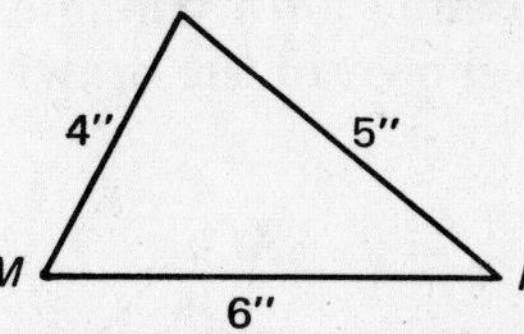


$\triangle MNP$ is *scalene.*

**Classification of Triangles by the Size of Their Angles**

**Acute triangle:** All three angles are *acute* (less than a right angle).

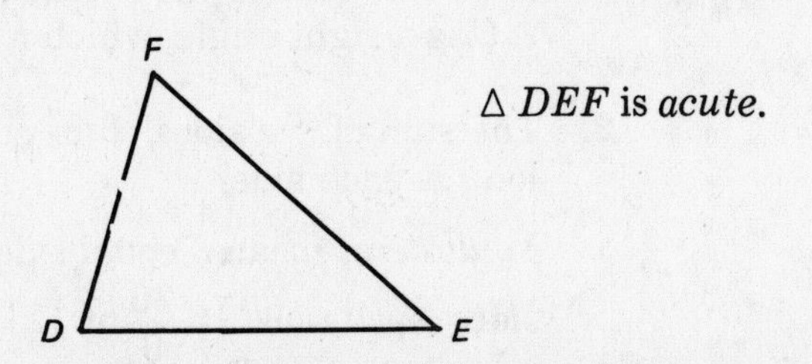


$\triangle DEF$ is *acute.*

**Obtuse triangle:** One angle is *obtuse* (greater than a right angle).

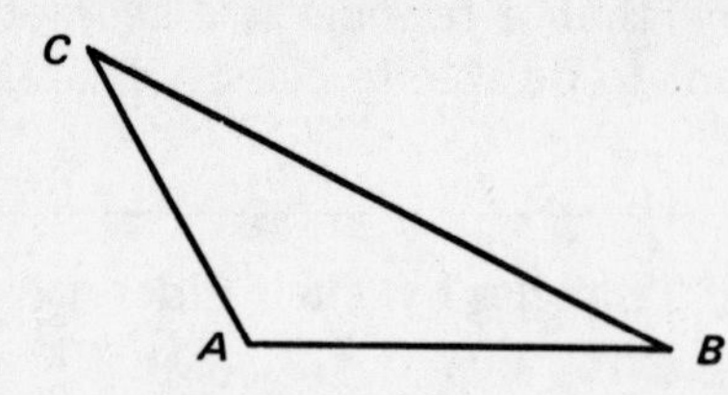


$\triangle$ *CAB* is obtuse, since
$\angle$ *A* is more than a right angle.

**Right triangle:** One angle is a *right* angle (90°).

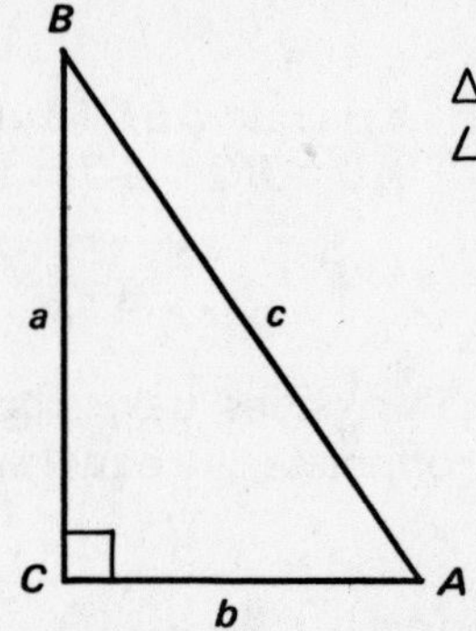


$\triangle$ *ABC* is a right triangle, since
$\angle$ *C* is a right angle (90°).

The longest side of a right triangle is called the <u>hypotenuse</u>. It is always the side opposite the right angle (side *c*). The other two sides are called *legs*.

*Examples:* 1. Is it possible for a triangle to be both *right* and *isosceles*? Yes, as shown in the drawing:

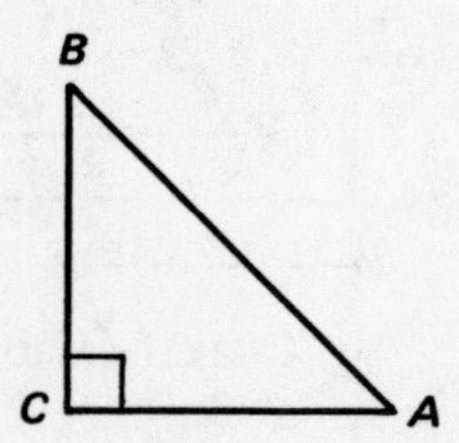


$BC = AC$ makes the triangle *isosceles*.
$\angle C$ is a right angle, which makes $\triangle ABC$ a *right triangle.*

2. The sum of the sides of an equilateral triangle is 36″. How long is each side?

*Equilateral* means "equal sided." Since there are three equal sides, each must be $\frac{36}{3}$ or 12″.

### The Pythagorean Theorem

There is a very important idea connected with the right triangle. It is known as the *Pythagorean Theorem* (or Pythagoras's Theorem). It says: "In a right triangle, the square of the hypotenuse is equal to the sum of the squares of the legs" ($a^2 + b^2 = c^2$).

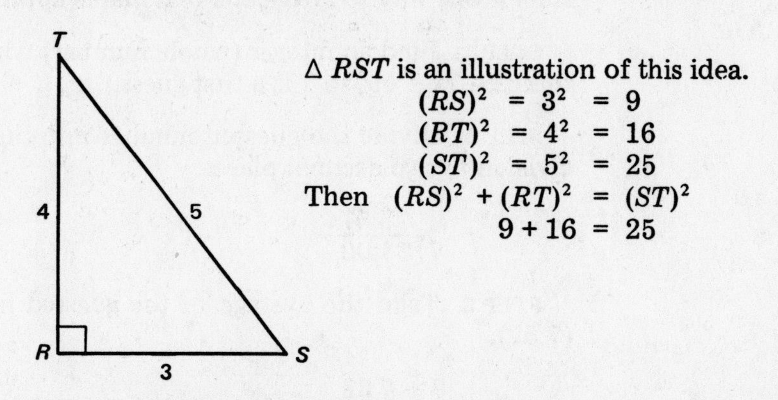


$\Delta\, RST$ is an illustration of this idea.

$$(RS)^2 = 3^2 = 9$$
$$(RT)^2 = 4^2 = 16$$
$$(ST)^2 = 5^2 = 25$$
$$\text{Then} \quad (RS)^2 + (RT)^2 = (ST)^2$$
$$9 + 16 = 25$$

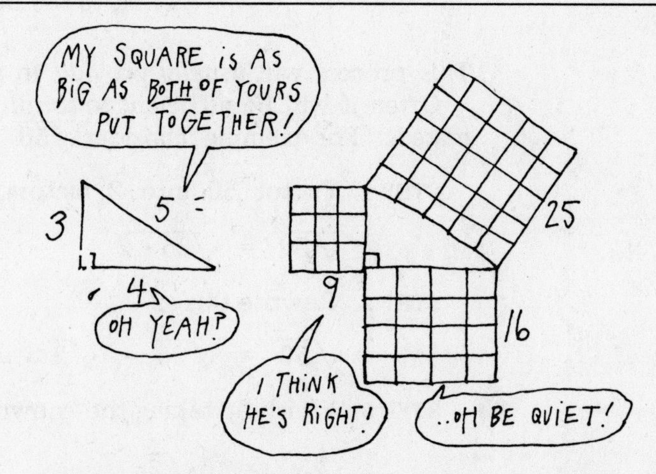


*Example*: If the hypotenuse of a right triangle is 13 and one leg is 12, find the other leg.

STEP 1. Draw and label a figure.

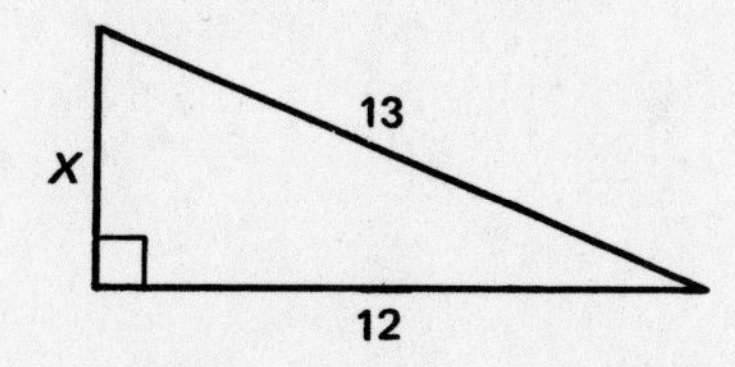


STEP 2. Set up an equation using the Pythagorean Theorem.

$$x^2 + 12^2 = 13^2$$
$$x^2 + 144 = 169$$
$$x^2 = 25$$
$$x = 5$$

Frequently you will find an equation similar to $x^2 = 25$. An equivalent form for this is: $x = \sqrt{25}$. The symbol $\sqrt{n}$ means "the *square root* of $n$" or "the number which multiplied by itself will equal $n$." The square root of 25 (or $\sqrt{25}$) is 5, and $\sqrt{49} = 7$.

More often, the square root of a number will not be a whole number, and in fact there will be no exact decimal representation of it.

Consider $\sqrt{53}$. Suppose we are to approximate this to the nearest tenth. Here is one way to arrive at a reasonable approximation.

STEP 1. Find an integer (whole number) whose square is *close to* 53. In this case $7^2 = 49$, so 7 is a first guess.

STEP 2. Divide the guessed number into the original number. Carry the division to two decimal places.

$$7\overline{)53.00} = 7.57$$

STEP 3. Take the average of the guessed number (7) and the quotient (7.57).

$$\frac{7 + 7.57}{2} = 7.23 \text{ or, to the nearest tenth, } 7.2$$

This process will usually provide an adequate method of approximation.

Often it will be sufficient to *simplify* a square root rather than approximate it. For example, consider $\sqrt{50}$.

STEP 1. Factor 50 into 2 factors, one of which is a perfect square.

$$\sqrt{50} = \sqrt{25 \cdot 2}$$

STEP 2. Rewrite this as

$$\sqrt{50} = \sqrt{25} \cdot \sqrt{2}$$

STEP 3. Finish by taking the known square root.

$$\sqrt{50} = 5\sqrt{2}$$

*Example*: If the legs of a right triangle are 7 and 8, find to the nearest tenth the length of the hypotenuse.

STEP 1. Draw and label a diagram.

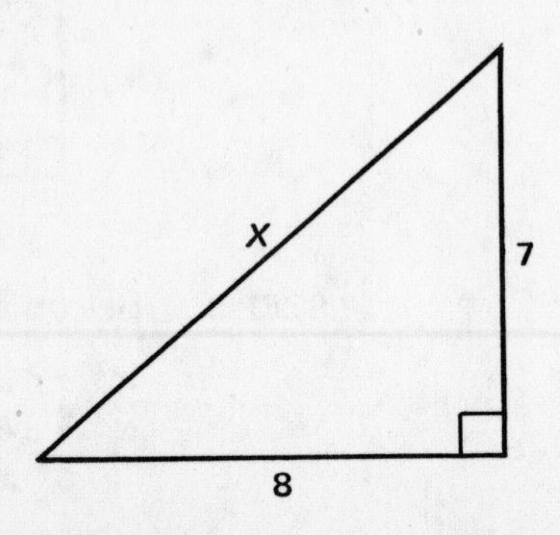

STEP 2. Set up an equation and solve.

$$
\begin{aligned}
7^2 + 8^2 &= x^2 \\
49 + 64 &= x^2 \\
113 &= x^2 \\
x^2 &= 113 \\
x &= \sqrt{113} \\
x &\doteq 10.6 \quad (\doteq \text{ means "approximately equal to"})
\end{aligned}
$$

CHECK $(10.6)^2 = 112.36$, which $\doteq 113$.

**Exercises** In a right triangle the hypotenuse is usually labeled $c$ and the legs are $a$ and $b$. Find the missing side in each of the following. If the answer is not exact, express it to the nearest tenth.

1. $a = 6, \quad b = 8, \quad c = ?$
2. $a = 7, \quad b = ?, \quad c = 25$
3. $a = ?, \quad b = 24, \quad c = 26$
4. $a = 5, \quad b = 6, \quad c = ?$
5. $a = 7, \quad b = ?, \quad c = 14$

**Answers** 1. 10 2. 24 3. 10 4. 7.8 5. 12.1

One other important idea about triangles concerns the sum of all the angles of any triangle. On the left in the picture below is a triangle. On the right the three angles of the triangle have been "torn off" and placed together.

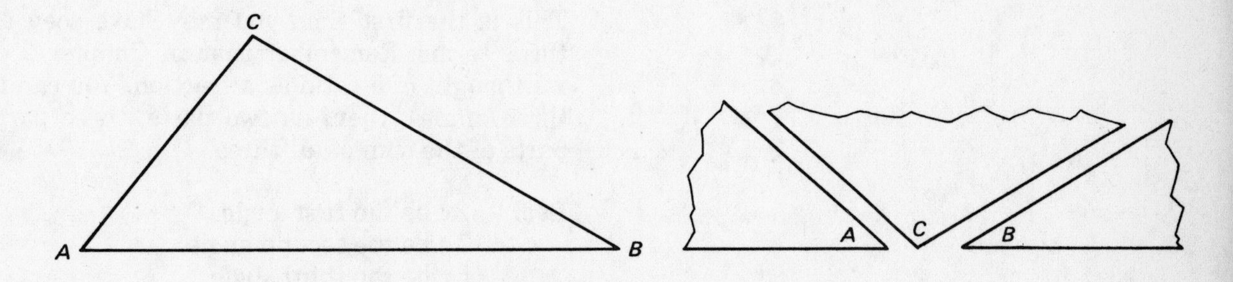


From this drawing it's easy to see that all three angles added together make a *straight angle*, or 180°.

| The sum of the three angles of any triangle is 180°. |
|---|

*Examples*: **1.** If one of the equal angles of an isosceles triangle is 40°, find each of the other two.

First draw a diagram.

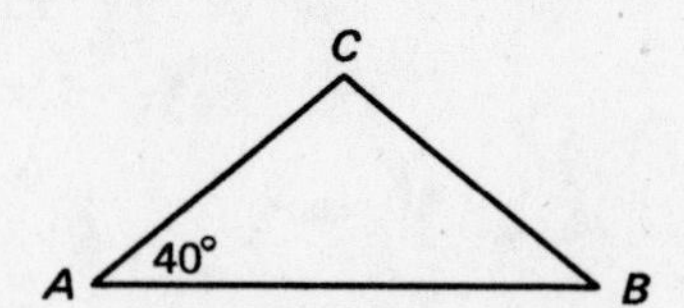


$\angle A = \triangle B$, therefore $\angle B$ is also 40°.
$\angle A + \angle B = 80°$. $\angle C$, then, must be 180° − 80°, or 100°. The angles are 40°, 40°, and 100°.

**2.** If one of the angles of a right triangle is 30°, what are the other two?

Again make a drawing, remembering that a right triangle contains a 90° angle.

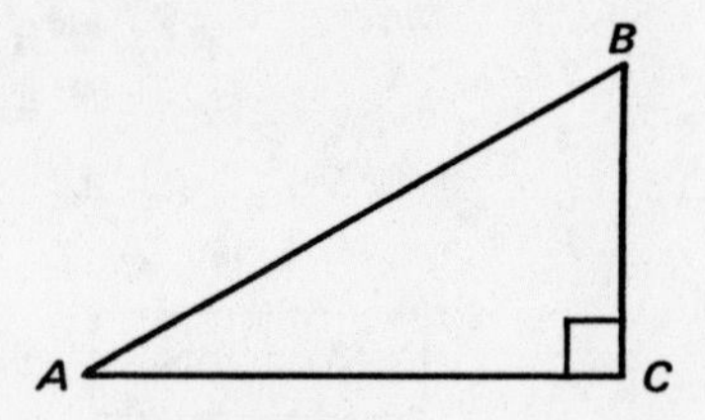


Since $\angle C = 90°$ and $\angle A = 30°$, $\angle B$ must be
180° − (90° + 30°) or 60°.

**3.** The angles of a triangle are in the ratio 2 : 3 : 5. Find the three angles.

This is the first time you may have seen a ratio with three terms. Remember that in Chapter 2 on fractions we thought of a ratio as a fraction. You can think of the three angles, then, as two parts, three parts, and five parts of the sum of all three.

Let: $2x$ be the first angle.
$3x$ be the second angle.
$5x$ be the third angle.

$2x + 3x + 5x = 180°$ is the equation.
$10x = 180°$
$x = 18°$

Then: $2x = 36°$ (first angle)
$3x = 54°$ (second angle)
$5x = 90°$ (third angle)

**Exercises**

1. Find all the angles of an isosceles right triangle.
2. Find all the angles of an equilateral triangle.
3. One of the equal angles of an isosceles triangle is 70°. Find the other two angles.
4. One angle of a triangle contains 40°. Find the other two angles if the second is four times the third.
5. If the angles of a triangle are in the ratio 2 : 5 : 8, what is the size of each angle?

**Answers**

1. 90°, 45°, 45°. Since one angle is 90°, the other two are each half of 90°.
2. 60°, 60°, 60°, since an equilateral triangle has three *equal* angles and 180° ÷ 3 = 60°.
3. 70°, 70°, 40°, as in example 1 about the isosceles triangle on page 130.
4. 40°, 28°, 112°

   Since 40° = first ∠,

   let: $x$ = third ∠

   $4x$ = second ∠

   $40 + x + 4x = 180°$

   $40 + 5x = 180°$

   $5x = 140°$

   $x = 28°$

   $4x = 112°$
5. 24°, 60°, 96°

   Let: $2x$ = first ∠

   $5x$ = second ∠

   $8x$ = third ∠

   $2x + 5x + 8x = 180°$

   $15x = 180°$

   $x = 12°$

   $2x = 24°$

   $5x = 60°$

   $8x = 96°$

# 8.4 Special Lines in a Triangle

Altitude: a line from a vertex that falls perpendicular to the opposite side, or to an *extension* of the opposite side.

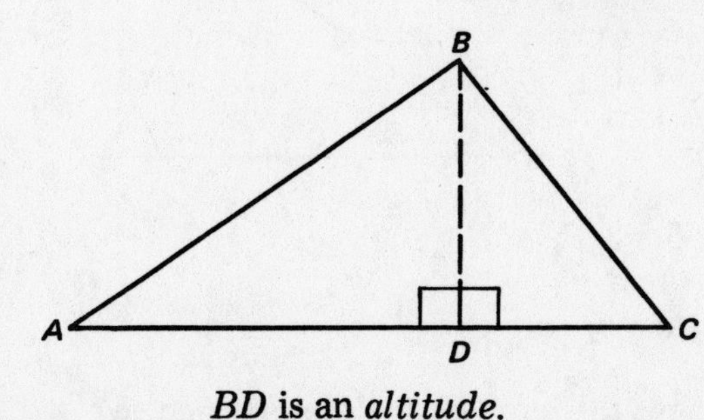


*BD* is an *altitude.*

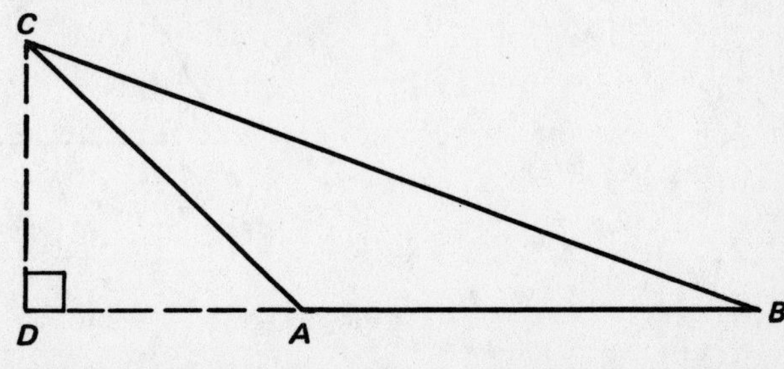


*CD* is an *altitude.* Notice that *AB* must be extended for the perpendicular to meet it at right angles.

**Median:** a line drawn from a vertex to the *middle point* of the opposite side.

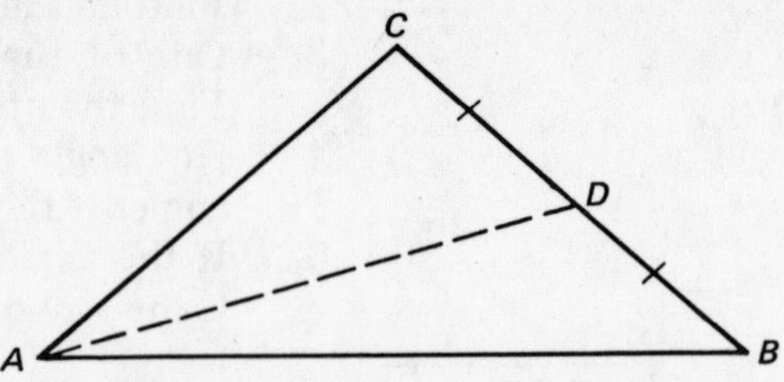


$AD$ is a *median* if $CD = DB$.

**Angle bisector:** a line drawn through a vertex so that it *bisects* the angle (divides it into *two equal parts*).

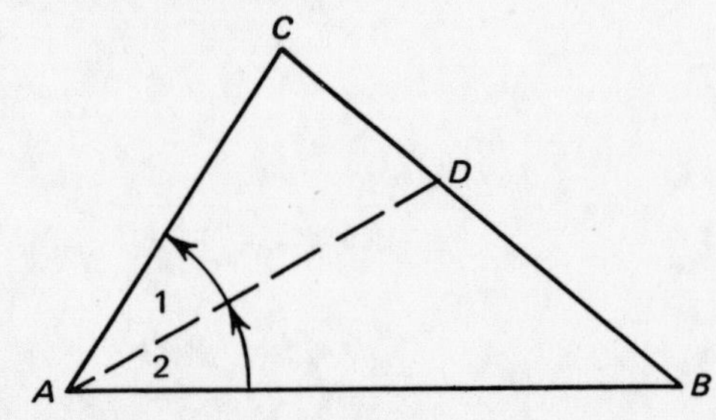


$AD$ is an *angle bisector of* $\angle A$ if $\angle 1 = \angle 2$.

# 8.5 Congruent Triangles and Similar Triangles

Special pairs of triangles fall into two categories, congruent and similar.

DEFINITION

**Congruent triangles** are triangles that are alike in all respects. *All three angles* and *all three sides* of one are *exactly the same* as those of the second.

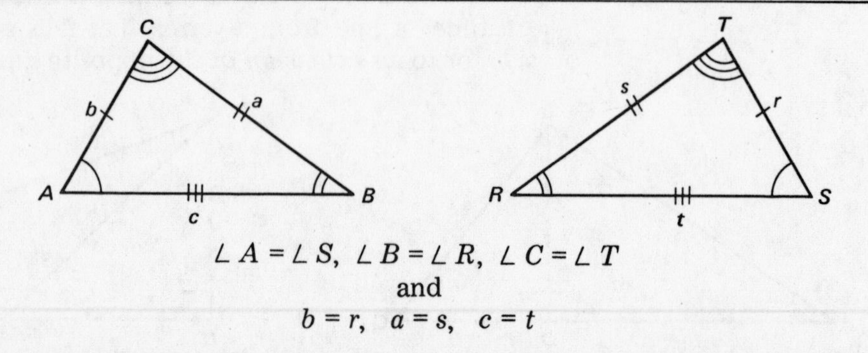


$$\angle A = \angle S, \ \angle B = \angle R, \ \angle C = \angle T$$

and

$$b = r, \ a = s, \ c = t$$

$\triangle ABC$ is congruent to $\triangle RST$ (using the symbol for congruence, we write this $\triangle ABC \cong \triangle RST$).

DEFINITION

**Similar triangles** are triangles which have the same shape, but not necessarily the same size. In order for two triangles to be similar, they must have *corresponding angles equal in measure* and *corresponding sides in the same ratio.*

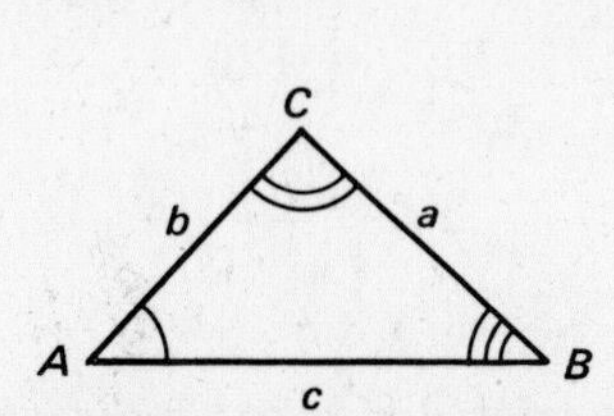


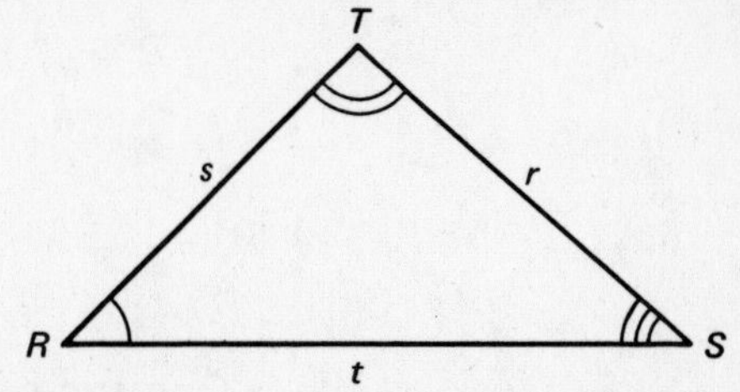


In the above illustration, in order for $\triangle ABC$ to be similar to $\triangle RST$ (symbolically, $\triangle ABC \sim \triangle RST$) it is necessary that:

$$\angle A = \angle R, \quad \angle B = \angle S, \quad \angle C = \angle T$$

and

$$\frac{b}{s} = \frac{a}{r} = \frac{c}{t}$$

NOTE: Congruent triangles are also similar, but similar triangles need not be congruent.

*Example*: The sides of one triangle are 5″, 12″, and 13″. Find the sides of a similar triangle whose shortest side is 10″.

STEP 1. Sketch the figure and label it.

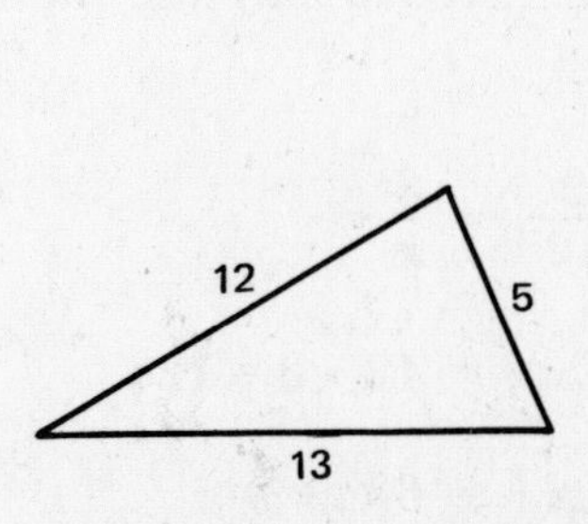


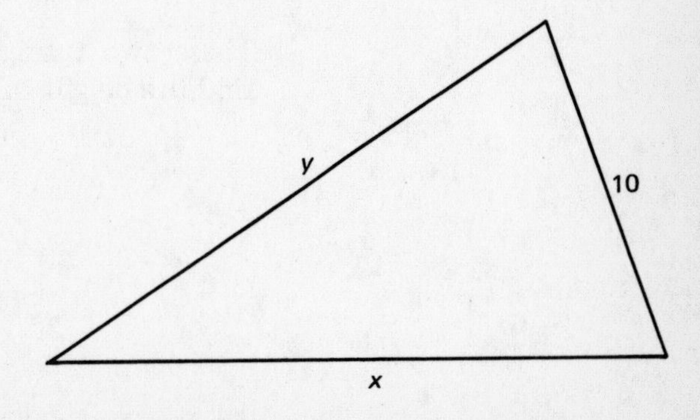

STEP 2. Set up equations that can be used to find the missing sides. Use the fact that the ratios of the sides must be the same.

$$\frac{5}{10} = \frac{12}{y} \qquad \frac{5}{10} = \frac{13}{x}$$

$$5y = 120 \qquad 5x = 130$$

$$y = 24 \qquad x = 26$$

One use of similar triangles is for *indirect* measurement. The meaning of indirect measurement will be made clear with an illustration.

*Example*: A vertical yardstick casts a shadow 2′ long. At the same time a tree casts a shadow 42′ long. Find the height of the tree.

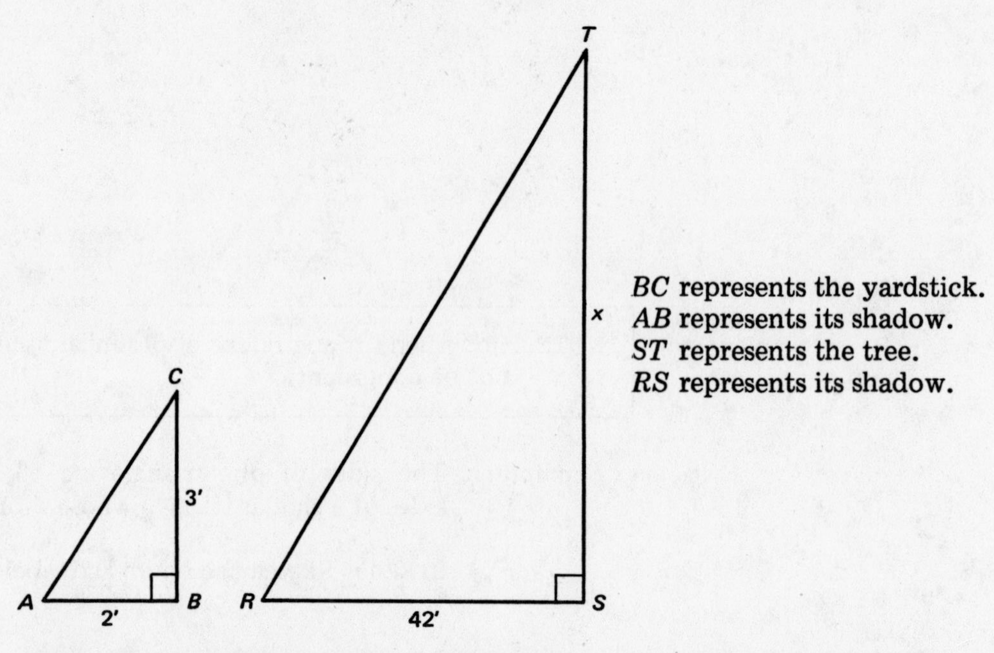


$BC$ represents the yardstick.
$AB$ represents its shadow.
$ST$ represents the tree.
$RS$ represents its shadow.

These two triangles are similar. We may use the idea of proportionality to find the height of the tree.

$$\frac{2}{42} = \frac{3}{x}$$

$$2x = 42 \cdot 3 = 126$$

$$x = 63'$$

# 8.6 Perimeter and Area

Before discussing these ideas, look at a new word: polygon.

DEFINITION

A **polygon** is composed of *three or more line segments* connected so that an area is closed in. The prefix *poly* means "many" and *gon* means "angle." So a polygon is a many-angled or many-sided figure.

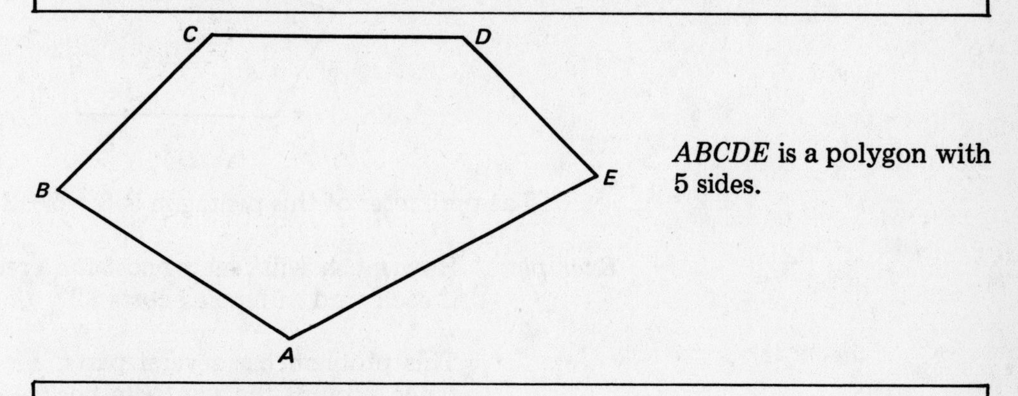


*ABCDE* is a polygon with 5 sides.

**Classification of Polygons**

| | |
|---|---|
| 3 sides | **triangle** (*tri* means 3) |
| 4 sides | **quadrilateral** (*quad* means 4) |
| 5 sides | **pentagon** (*pent* means 5) |
| 6 sides | **hexagon** (*hex* means 6) |
| 8 sides | **octagon** (*oct* means 8) |
| 10 sides | **decagon** (*dec* means 10) |

There are several distinct types of quadrilateral that you should understand before attempting problems. Here is a brief description of them:

DEFINITIONS

A **parallelogram** is a quadrilateral with its opposite sides parallel. In a parallelogram, the opposite sides and angles are also equal.

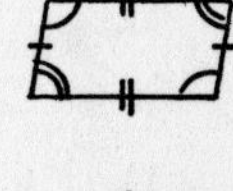

A **rectangle** is a parallelogram in which all the angles are right angles.

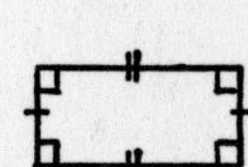

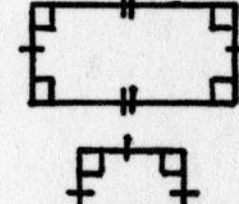

A **square** is a rectangle with all four sides equal.

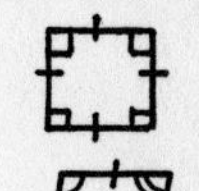

A **rhombus** is a parallelogram in which all four sides are equal. Notice that a square is a rhombus, but that a rhombus is not necessarily a square.

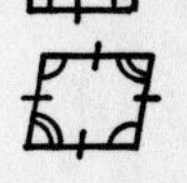

A **trapezoid** is a quadrilateral with two sides parallel and two sides not parallel.

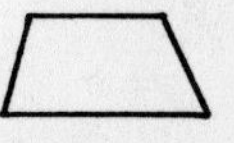

Now we can move on to the concept of *perimeter.*

DEFINITION

The **perimeter** of a polygon is *the sum of all its sides.*

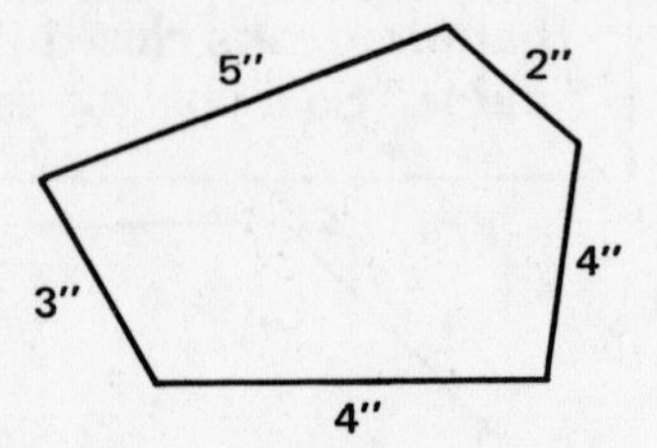


The perimeter of this pentagon is 3 + 5 + 2 + 4 + 4 = 18".

*Example*: How much will fencing cost for a rectangular yard 42′ by 60′ if each yard of fencing costs $2?

This problem has several parts. First we must find the distance around the yard (its *perimeter*). A picture will help.

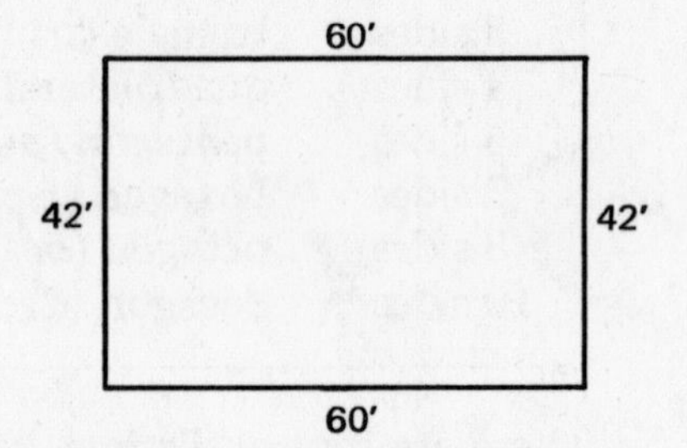


The perimeter is 204′ (42′ + 60′ + 60′ + 42′). This is the number of *feet.* Since there are 3′ in one yard, divide 204′ by 3 to get 68 yards. Now it's easy to find the cost, since 68 × $2 = $136.

Exercises

Try the following problems to make sure you understand the idea of *perimeter*.

1. An equilateral pentagon has a perimeter of 75". Find the length of each side.
2. In rectangle *ABCD,* a diagonal line (*AC*) is drawn. If *AB* = 4" and *BC* = 3", what is the perimeter of $\triangle ABC$?
3. If molding costs $0.12 a foot, what will be the cost of molding for a living room which is 12′ by 18′?

**Answers**

1. 15″ Just divide 75 by 5.
2. 12″ Draw this picture:

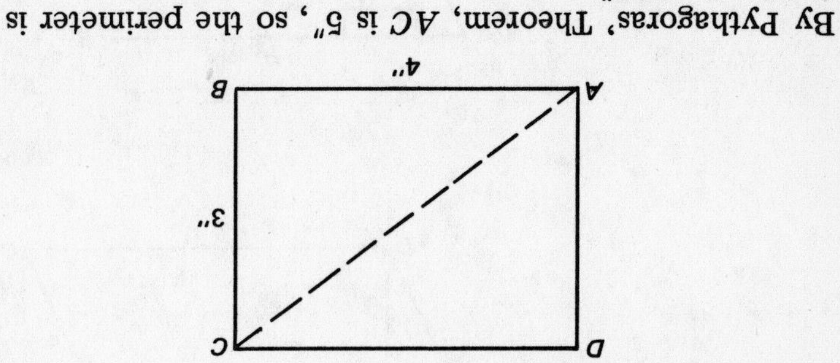


By Pythagoras' Theorem, $AC$ is 5″, so the perimeter is 3 + 4 + 5 or 12″.

3. \$7.20 The living room has a perimeter of 60′ (12 + 18 + 12 + 18). 60 × \$.12 = \$7.20 for the molding.

Now we will deal with the concept of *area.*

DEFINITION

**Area** is the space enclosed by a polygon.

To figure the *area* of a polygon, look first at a rectangle.

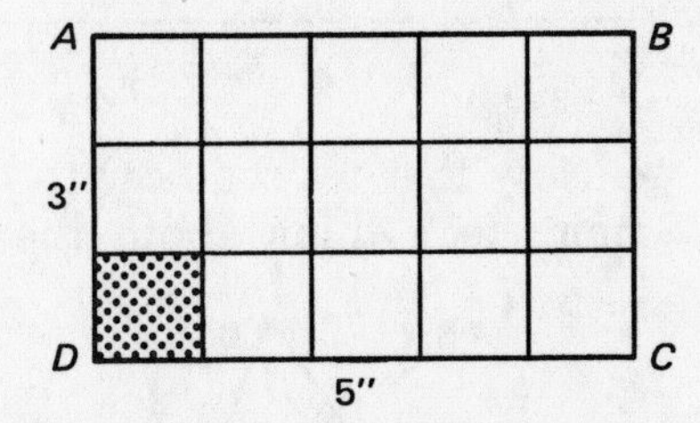


If the rectangle is 3″ by 5″, it will take 15 little squares like the one in the lower left corner to cover it. These little squares, 1″ on a side, are *square inches.* A square is the basic unit for measuring area.

*For any rectangle, the area is the product of the length and width.* (In the above diagram, $DC$ may also be called the *base* of the rectangle. $AD$ may be called the *altitude* or *height* of the rectangle.) Then the area of the above triangle is 5 × 3 or 15 square inches.

The formula for the area of a rectangle:
$A = bh$ ($b \times h$, or base × altitude)

Now let's take a look at some other geometric shapes, and then find their areas. A *parallelogram* is a four-sided polygon whose opposite sides are parallel. The formula for the area of a parallelogram is the same as that for the area of a rectangle.

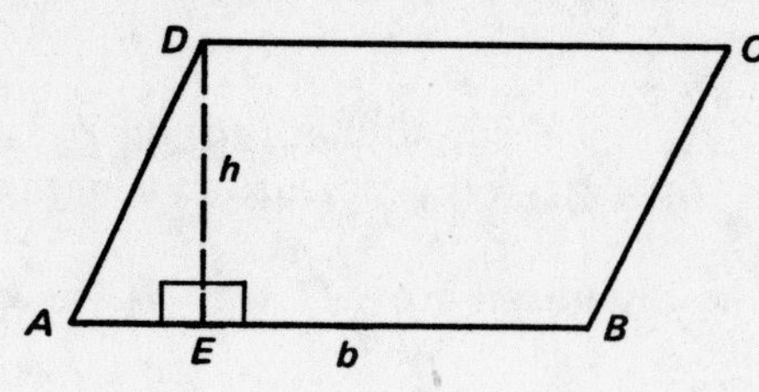


$ABCD$ is a parallelogram.
$DE \perp AB$
$AB$ is the base.
$DE$ is the altitude.

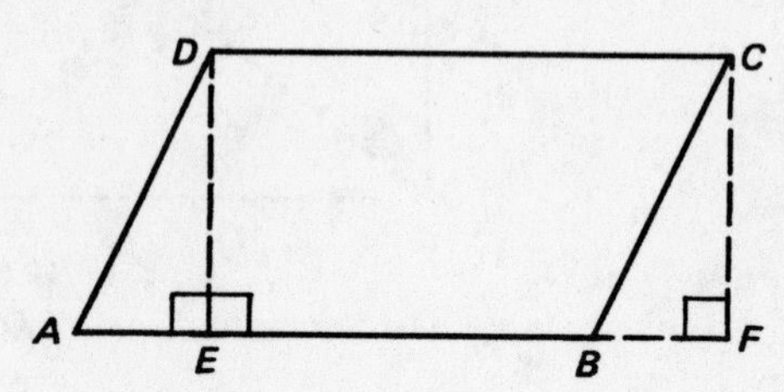


This is the same parallelogram. If $\Delta\ ADE$ is cut off and placed where $\Delta\ BCF$ is, the new figure is $EDCF$. This is a rectangle with the same area as parallelogram $ABCD$. Its area formula is the same also: $A = b \times h$.

The formula for the area of a parallelogram:
$A = bh$ (base $\times$ altitude)

CAUTION: Do not confuse the side of a parallelogram with its altitude.

Now that we see how to compute the area of a rectangle and a parallelogram, let's think about the problem of computing the area of a triangle.

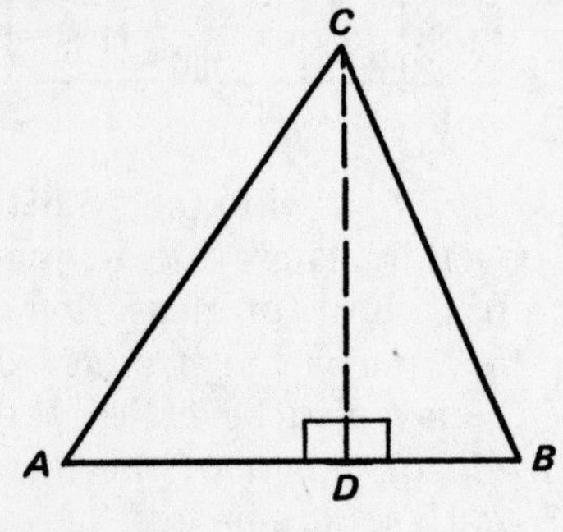


$AB$ is the *base* of the triangle.
$CD$ is the *altitude* or height of the triangle.

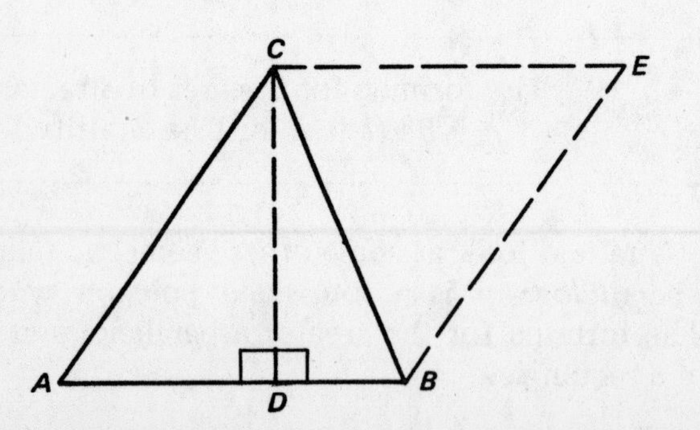


In this figure, $ABC$ is the same triangle. We construct another triangle, $BCE$, congruent to $ABC$. The resulting figure, $ACEB$, is a parallelogram with *base* $AB$ and *altitude* $CD$. The area of $\Delta\ ABC$ is $\frac{1}{2}$ the area of the parallelogram or $\frac{1}{2}\ b \times h$.

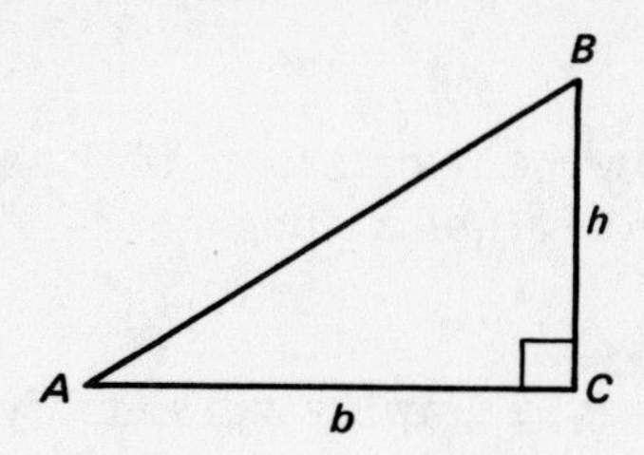


SPECIAL CASE: In the *right* triangle $ABC$, either leg $AC$ or leg $BC$ is the base, and the other leg is the altitude.

The formula for the area of a triangle:

$$A = \frac{1}{2}\,bh \quad \text{or} \quad \frac{bh}{2}$$

Problems about area depend on one or more of the formulas plus some use of algebra.

*Examples*: 1. A parallelogram and a square have equal areas. The base of the parallelogram is 3 more than a side of the square. Its altitude is 2 less than the side of the square. Find a side of the square. (REMEMBER: All sides of a square are equal.)

Start with a diagram of each figure:

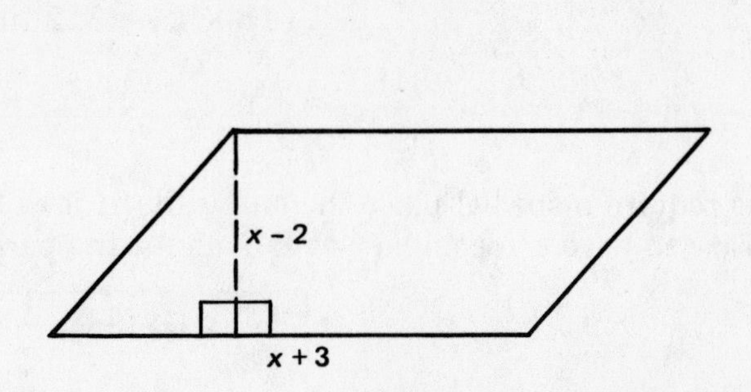


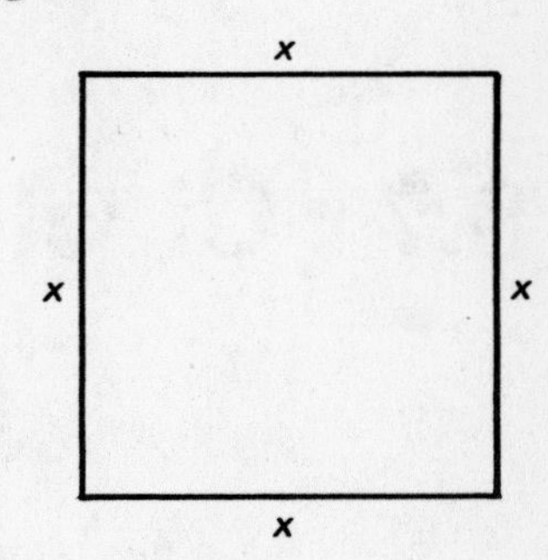


Label the side of the square $x$.
From the problem, the base of the parallelogram is $x + 3$ and its altitude is $x - 2$.
The area of the square is $x \cdot x$, or $x^2$.
The area of the parallelogram is $(x + 3)(x - 2)$.
Since the two areas are equal, the equation is:

$$\begin{aligned}
(x + 3)(x - 2) &= x^2 \\
x^2 + x - 6 &= x^2 \\
x - 6 &= 0 \\
x &= 6 \text{ (the side of the square)} \\
(6) + 3 &= 9 \text{ (the base of the parallelogram)} \\
(6) - 2 &= 4 \text{ (the altitude of the parallelogram)}
\end{aligned}$$

As a check, you can see that the area of the square is $6 \times 6$ or 36, and that the area of the parallelogram is $9 \times 4$, which also is 36.

2. How many bathroom tiles, each 4″ on a side, are needed to tile a wall 8′ long and 4′ 4″ high?

Again, a drawing will simplify the job:

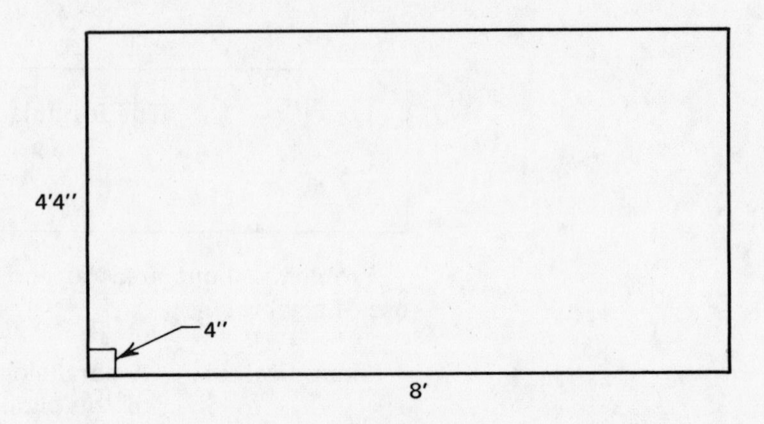


There is a trap in this problem: the units of measure are not the same. To stretch across the bottom of the rectangle 24 tiles are needed (8 feet = 96 inches. 96 ÷ 4 = 24 tiles). One row *up* will use 13 tiles (4′4″ = 52″. 52 ÷ 4 = 13 tiles).

$13 \times 24 = 312$ tiles needed.

# 8.7 Circles

Circles require a special place in our study of area and perimeter. First, it's necessary to have a vocabulary to talk about the parts of a circle.

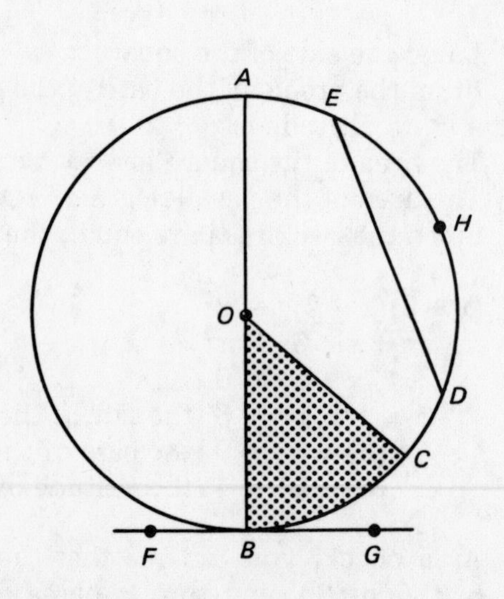


**Special Lines in a Circle**

*OC* is a **radius**. It is a line segment drawn from the center to any point on the circle.

*ED* is a **chord**. It is a line joining any two points on the circle.

*AB* is a **diameter**. It is a chord that passes through the center of the circle, and equals *twice the radius*.

*FG* is a **tangent** to the circle. It is a line that has *only one point* in common with the circle.

*EHD* is an **arc**. This is the portion of the circle lying between two points.

*ADB* is a **semicircle**. It is an arc also, and is exactly half the circle.

The region $OCB$ is a **sector**. It is an area bounded by two radii and an arc.

$\angle COB$ is a **central angle**. It is formed by two radii.

The **circumference** of the circle is its length (exactly once around the "rim"). It encompasses 360°.

To compute the circumference ($C$) of a circle, you need a new number, $\pi$. The symbol $\pi$ is actually the Greek letter *pi*, which has had a special meaning in geometry for more than 2,000 years. Pi is the unchanging ratio of the circumference of a circle to its diameter: $\pi = \frac{C}{d}$. This would indicate that $C = \pi d$, or that the circumference of a circle is equal to $\pi$ times the diameter. The number $\pi$ is approximately equal to 3.14 or $3\frac{1}{7}$. The area of a circle is equal to $\pi$ times the square of the radius.

In a circle:

$$A = \pi r^2$$

$$C = \pi d \text{ or } C = 2\pi r$$

$$\pi \doteq 3.14 \text{ or } 3\frac{1}{7}$$

*Examples:*

1. Find the circumference and area of a circle with radius 10 inches. (Use $\pi \doteq 3.14$.)

   $C = \pi d\,(d = 2r)$

   $C = (3.14)(20) = 62.8$ inches

   $A = \pi r^2$

   $A = 3.14 \times 10^2 = 314$ square inches.

2. Find the area of a sector of a circle of radius 14″ if the central angle of the sector is 45°. (Use $\pi \doteq 3\frac{1}{7}$.)

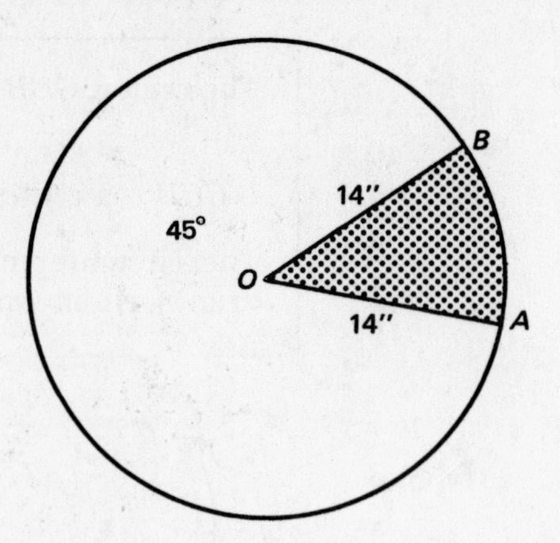


The shaded area *OAB* is the sector. The area of the entire circle is:

$A = \pi r^2$

$A = 3\frac{1}{7} \times 14^2$

$A = \frac{22}{7} \times 14 \times 14 = 616$ square inches.

The portion of the circle contained in the sector is $\frac{45}{360}$ or $\frac{1}{8}$ of the whole circle, since there are 360° in the whole circle.

$\frac{1}{8} \times 616 = 77$ square inches.

**3.** If the largest possible circular piece is cut from a 2′ square piece of wood, how much is wasted? (Use $\pi \doteq 3.14$.)

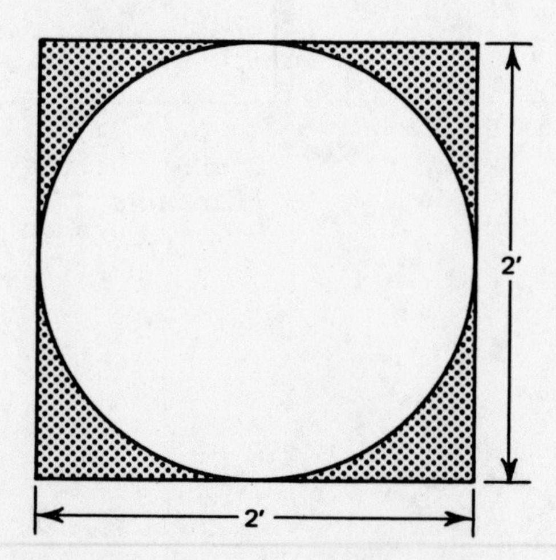


The shaded area represents the amount of waste.

STEP 1. Find the area of the square: $2 \times 2 = 4$ square feet.

STEP 2. Find the area of the circle: $3.14 \times 1^2$, or 3.14 square feet. (The radius must be half the diameter, or 1′.)

STEP 3. Subtract 3.14 from 4 to find the waste area: 0.86 square feet.

# 8.8 Solid Geometry

Three-dimensional or solid geometry is the study of the surface areas and volumes of solid figures.

> **DEFINITION**
>
> **Volume** means the space occupied by a solid figure.

While *length* is measured in *linear* units, and *area* is measured in *square* units, *volume* is measured in *cubic* units.

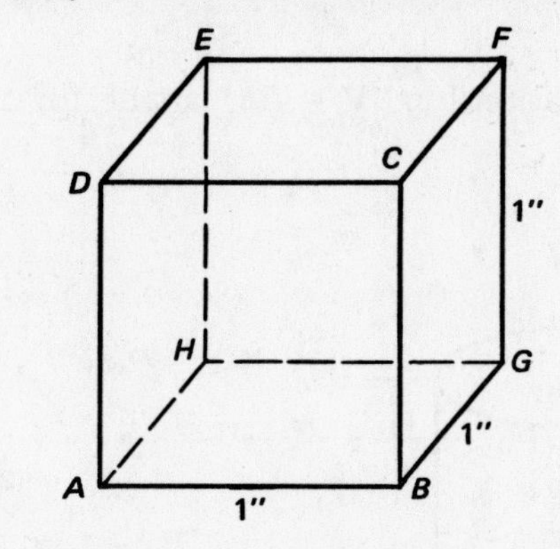


The figure at the left is a *cube* with all of its edges 1 inch. Its volume is 1 *cubic inch* ($1 \times 1 \times 1$). Square *ABCD* is called a *face*. There are 6 faces on the cube.

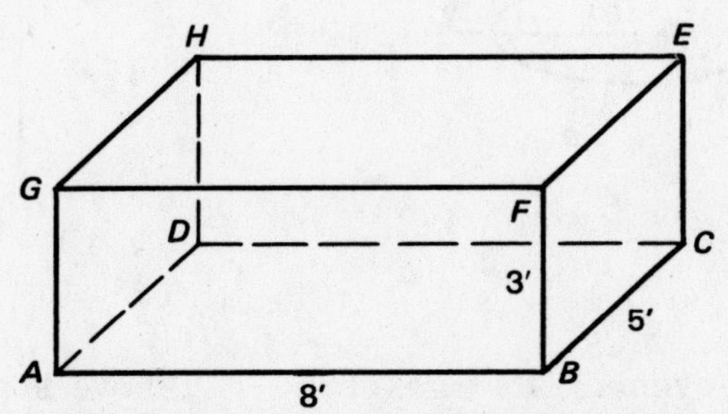


The figure at the left is a *rectangular solid*. Rectangle *ABCD* is its *base*. Any vertical edge, such as *FB* or *HD*, is its *altitude*.

Two problems to consider are the surface area and the volume of the solid above.

*Example*: If $AB = 8'$, $BC = 5'$ and $BF = 3'$, find the *surface area* and *volume* of the figure.

**Surface area:** There are 6 separate rectangles in the solid. Find the area of each and add.

$ABCD = 8 \times 5$ or 40 square feet
$GFEH = 8 \times 5$ or 40 square feet
$BCEF = 5 \times 3$ or 15 square feet
$ADHG = 5 \times 3$ or 15 square feet
$ABFG = 8 \times 3$ or 24 square feet
$DCEH = 8 \times 3$ or 24 square feet

Surface area = 158 square feet

**Volume:** Imagine a layer of 1-foot cubes on the base of the solid. There should be $8 \times 5$ or 40 of them to cover the entire base. This is the same as the area of the base. Since the height is $3'$, three layers of such cubes would be needed to completely fill the solid.

Volume = $3 \times 40$ or 120 *cubic* feet

| Volume of a rectangular solid: | $V = Bh$<br>$B$ is the *area* of the base.<br>$h$ is the height or altitude. |
|---|---|

This same relationship, $V = Bh$, is true for cylinders and prisms as well.

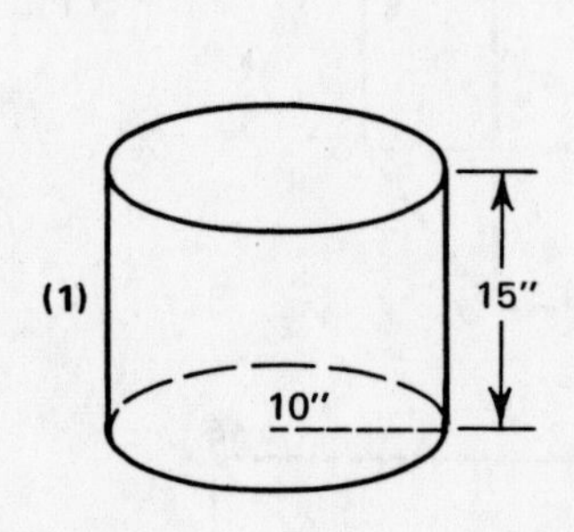


Above is a *cylinder*. Its bases are circles, and in this case its altitude is $15''$.

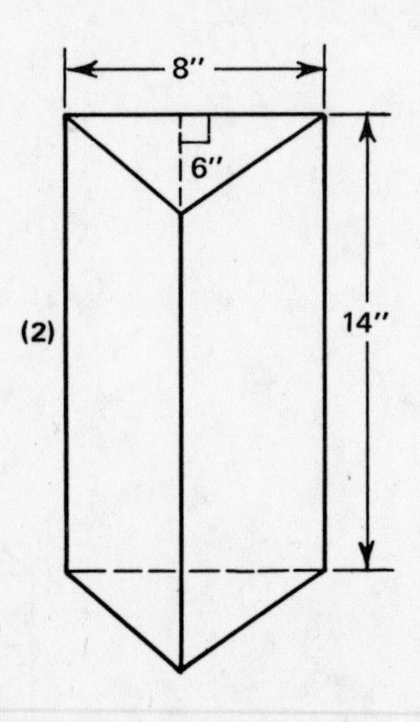


Above is a triangular *prism*. It is similar to the rectangular solid, except its bases are triangles.

(1) Volume of a cylinder

To compute the volume, use the formula: $V = Bh$. Here $B$ is the base, or a circle with radius 10″, and $h = 15″$.

$B = \pi r^2$

$B = 3.14 \times 10^2$

$B = 314$ square inches

$V = B \times h$

$V = 314 \times 15$

$V = 4{,}710$ cubic inches

(2) Volume of a prism

To compute the volume, use the same formula: $V = Bh$. Here the base $B$ is a *triangle* with base $b = 8″$ and altitude $h = 6″$.

$B = \frac{1}{2} bh$

$B = \frac{1}{2} \times 8 \times 6$

$B = 24$ square inches

$V = Bh$

$V = 24 \times 14$

$V = 336$ cubic inches

# 8.9 Table of Formulas

AREA

|  |  |
|---|---|
| Rectangle | $A = lw$ |
| Square | $A = s^2$ |
| Parallelogram | $A = bh$ |
| Triangle | $A = \frac{1}{2} bh$ |
| Circle | $A = \pi r^2$ |

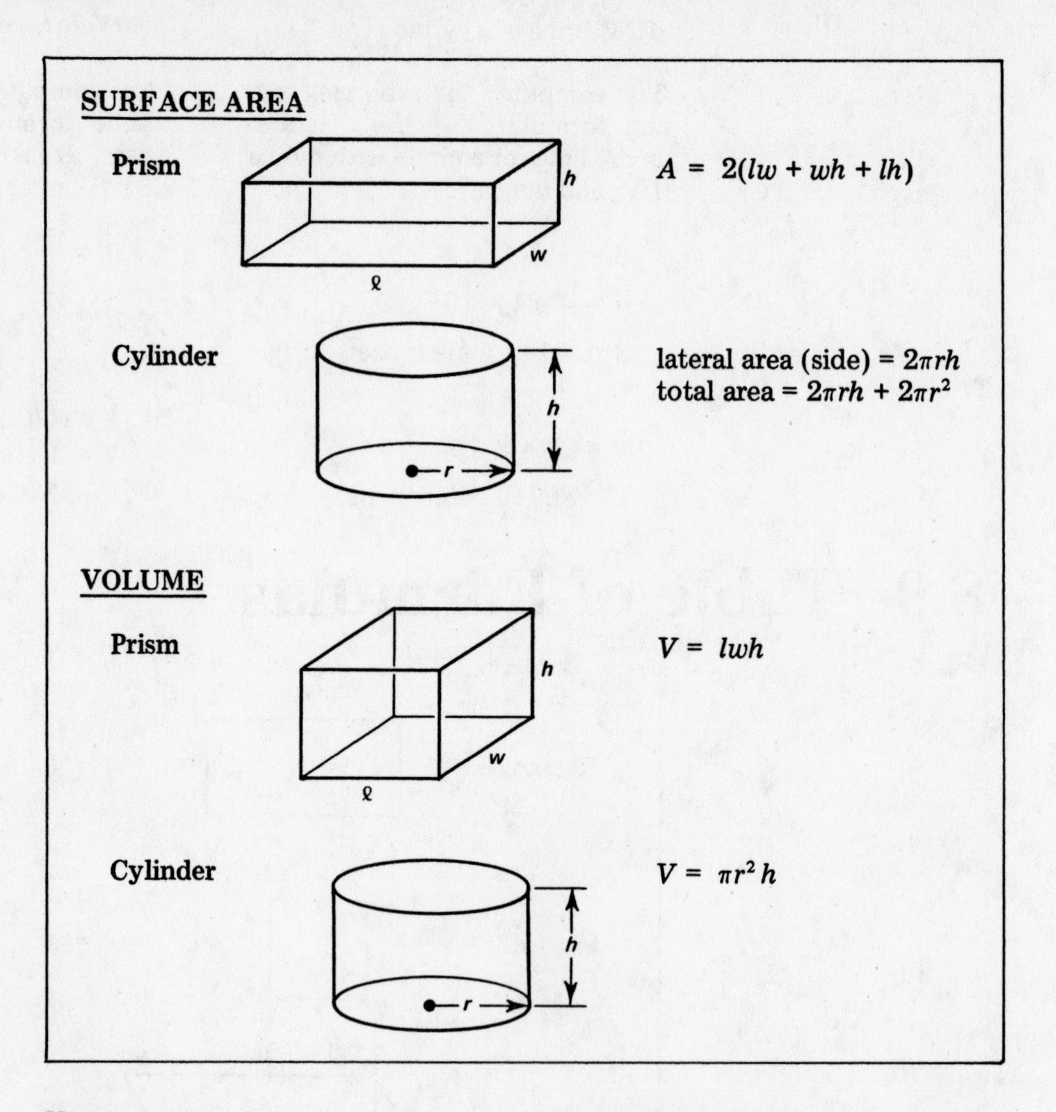


Here are some formulas that you will find useful in solving more advanced problems.

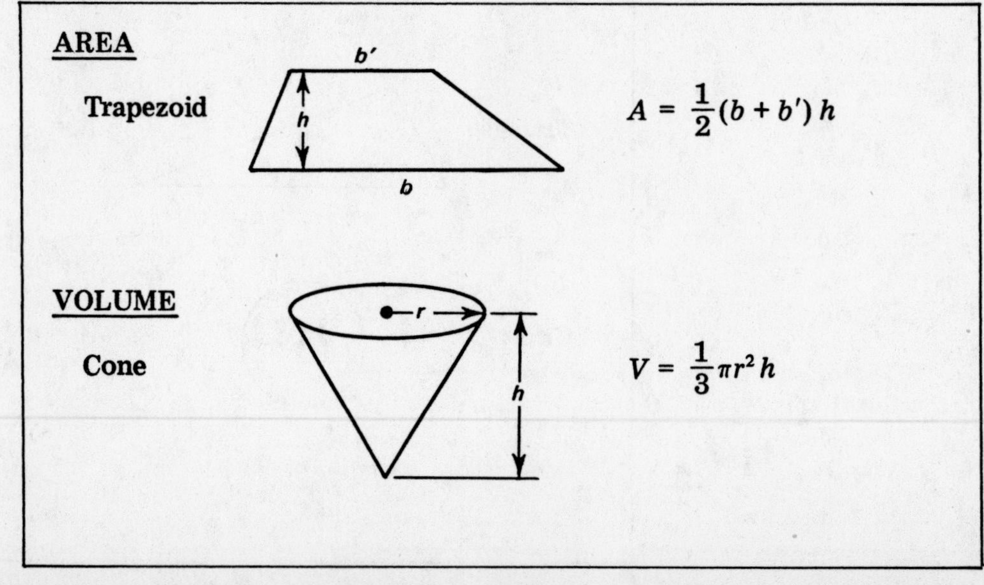

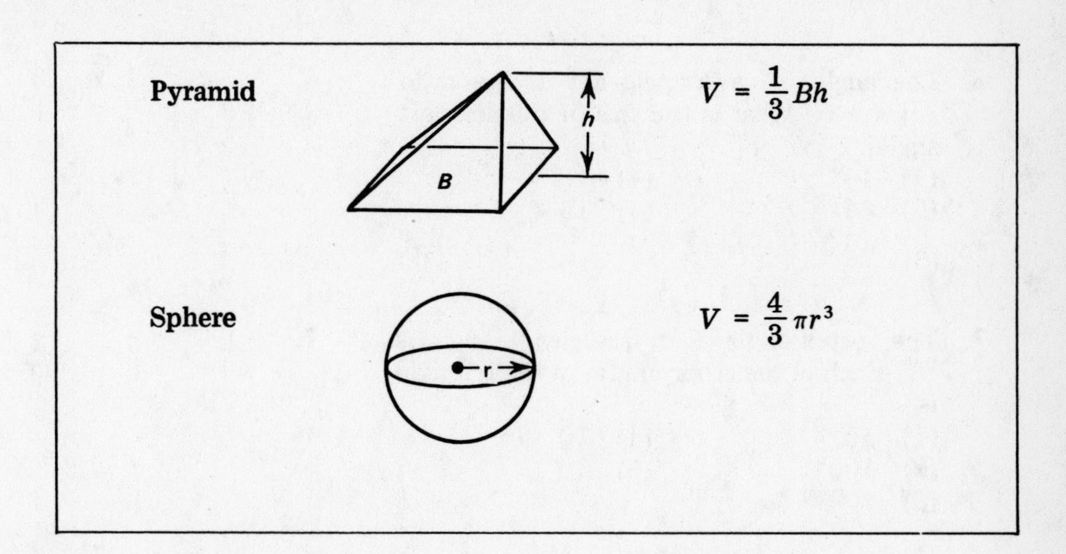


# Trial Test — Geometry

This is your achievement test on geometry. Darken the square under the letter of the correct answer.

1. What is the complement of an angle of 41°?
   (1) 41°
   (2) 49°
   (3) 139°
   (4) 90°
   (5) 180°

   1. 1 2 3 4 5

2. What is the supplement of an angle of 37°?
   (1) 37°
   (2) 53°
   (3) 143°
   (4) 153°
   (5) 133°

   2. 1 2 3 4 5

3. An isosceles triangle is a triangle with
   (1) two equal sides
   (2) no sides equal
   (3) three equal angles
   (4) three equal sides
   (5) no obtuse angle

   3. 1 2 3 4 5

4. Two angles of a triangle are 30° and 45°. How large is the third angle?
   (1) 105°
   (2) 15°
   (3) 75°
   (4) 95°
   (5) 60°

   4. 1 2 3 4 5

5. An angle whose measure is greater than 90° and less than 180° is called
   (1) acute
   (2) obtuse
   (3) right
   (4) straight
   (5) reflex

   5. 1 2 3 4 5

6. The angles of a triangle are in the ratio 1 : 4 : 7. What is the size of the smallest angle?
(1) 45°
(2) 30°
(3) $7\frac{1}{2}°$
(4) 18°
(5) 15°

6. 1 || 2 || 3 || 4 || 5 ||

7. The vertex angle of an isosceles triangle is 80°. Each of the other angles of the triangle is
(1) 50°
(2) 100°
(3) 20°
(4) 10°
(5) 40°

7. 1 || 2 || 3 || 4 || 5 ||

8. In $\triangle ABC$, $\angle C = 90°$. If $AC = 6''$ and $BC = 8''$ then $AB =$
(1) 2″
(2) 9.2″
(3) 8″
(4) 10″
(5) 9.6″

8. 1 || 2 || 3 || 4 || 5 ||

9. Find to the nearest tenth of an inch the hypotenuse of a right triangle whose legs are 5″ and 8″.
(1) 9.1″
(2) 9.2″
(3) 9.4″
(4) 9.5″
(5) 9.6″

9. 1 || 2 || 3 || 4 || 5 ||

10. The three sides of a triangle are 3, 5, and 6. Find the longest side of a similar triangle whose shortest side is 2.
(1) 8
(2) 3
(3) 5
(4) 6
(5) 4

10. 1 || 2 || 3 || 4 || 5 ||

11. A polygon with 6 sides is called a(n)
(1) quadrilateral
(2) pentagon
(3) hexagon
(4) octagon
(5) decagon

11. 1 || 2 || 3 || 4 || 5 ||

12. A circle has a radius of 21″. Its circumference is
(1) 1,386″
(2) 66″
(3) 132″
(4) 441″
(5) 142″

12. 1 || 2 || 3 || 4 || 5 ||

13. A right triangle has sides 5, 12, and 13. Find the area.
(1) 60
(2) 30
(3) 78
(4) 65
(5) $32\frac{1}{2}$

13. 1 || 2 || 3 || 4 || 5 ||

14. A rectangle and a square have the same area. If the length and width of the rectangle are 16′ and 9′, find the side of the square.
(1) 13′
(2) 10′
(3) 14′
(4) 12′
(5) 15′

14. 1 2 3 4 5

15. Two circles have the same center. If their radii are 7″ and 10″, find the area included between them.
(1) 3 sq. in.
(2) 9 sq. in.
(3) 9 $\pi$ sq. in.
(4) 51 $\pi$ sq. in.
(5) 51 sq. in.

15. 1 2 3 4 5

16. A box measures 24″ by 12″ by 9″. How many cubic inches does it contain?
(1) 45
(2) 324
(3) 2,592
(4) 2,692
(5) 334

16. 1 2 3 4 5

17. There are 231 cubic inches in a gallon. A cylindrical can has a radius of 7″ and a height of 15″. How many gallons does it contain?
(1) 15
(2) 12
(3) 13
(4) 14
(5) 10

17. 1 2 3 4 5

18. A patio 9′ by 12′ is to be 6″ thick. How many cubic yards of cement are needed to construct it?
(1) 2
(2) 648
(3) 54
(4) 72
(5) 3

18. 1 2 3 4 5

19. How many square tiles 9″ on a side will be needed to tile a floor 9′ by 12′?
(1) 108
(2) 192
(3) 144
(4) 156
(5) 216

19. 1 2 3 4 5

20. A pie is 9″ in diameter. What is the top area of a piece of pie making an angle of 45° at the center? Answer to the nearest square inch.
(1) 4
(2) 5
(3) 6
(4) 7
(5) 8

20. 1 2 3 4 5

21. If there are 100 centimeters in one meter, how many *square* centimeters are contained in a *square* meter?
(1) 10
(2) 100
(3) 1,000
(4) 10,000
(5) 1,000,000

21. 1 || 2 || 3 || 4 || 5 ||

22. Through a 5″ cube, a hole 1″ in radius is drilled. Find the volume of the resulting figure (to the nearest cubic inch).
(1) 16
(2) 9
(3) 125
(4) 109
(5) 91

22. 1 || 2 || 3 || 4 || 5 ||

23. A baseball diamond is 90′ (90′ on each side, *not* 90 sq. ft.). How many feet, to the nearest foot, is it from home plate to second base?
(1) 180
(2) 126
(3) 115
(4) 90
(5) 127

23. 1 || 2 || 3 || 4 || 5 ||

24. The earth is about 8,000 miles in diameter. Approximately how long is the equator?
(1) 16,000 miles
(2) 24,000 miles
(3) 25,000 miles
(4) 200,000 miles
(5) 1,000,000 miles

24. 1 || 2 || 3 || 4 || 5 ||

25. The base of a triangle is 3″ more than the side of the square. The height of the triangle is 2″ more than the side of the square. Find the side of the square, if each figure has the same area.
(1) 5
(2) 6
(3) 7
(4) 8
(5) 9

25. 1 || 2 || 3 || 4 || 5 ||

---

## SOLUTIONS

1. 2 Complementary angles have a sum of 90°.
2. 3 Supplementary angles have a sum of 180°.
3. 1 Isosceles means "having two equal sides."
4. 1 The sum of the angles of a triangle is 180°.

   $30° + 45° + x = 180°$. $x = 105°$.
5. 2
6. 5 Call the angles $x$, $4x$, $7x$.

   Then: $x + 4x + 7x = 180°$

   $12x = 180°$

   $x = 15°$
7. 1 Call each base angle $x$.

   Then: $x + x + 80 = 180°$

   $2x = 100°$

   $x = 50°$

8. 4 Use a diagram:

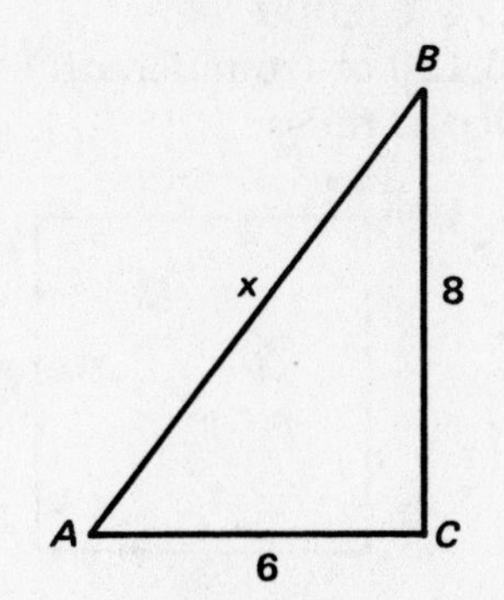


$x^2 = 6^2 + 8^2$
$x^2 = 36 + 64$
$x^2 = 100$
$x = 10$

9. 3 Use $c^2 = a^2 + b^2$. $c^2 = 5^2 + 8^2$
$c^2 = 89$
$c \doteq 9.4$

10. 5 Use proportionality of sides of similar triangles.

$\frac{3}{2} = \frac{6}{x}$

$3x = 12$

$x = 4$

11. 3

12. 3 Use $C = \pi d$. $C = \frac{22}{7} \times 42 = 132$

13. 2

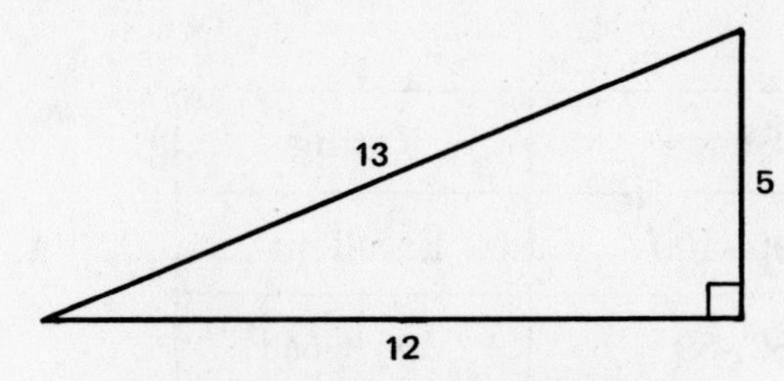


5 is the height and 12 is the base.

$A = \frac{1}{2} bh$

$A = \frac{1}{2} \times 12 \times 5 = 30$

14. 4 The area of the rectangle is 144 sq. ft. The area of the square is $x^2$.

$x^2 = 144$
$x = 12$

15. 4

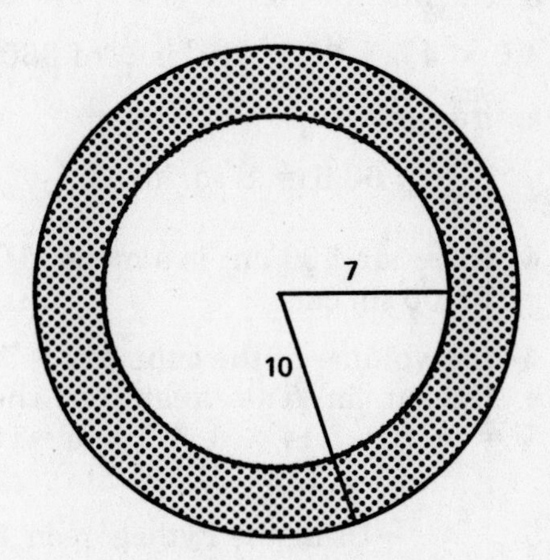


The area of the outer circle is:

$A = \pi r^2$
$A = \pi \times 10^2$
$A = 100\pi$

The area of the inner circle is:

$A = \pi r^2$
$A = \pi \times 7^2$
$A = 49\pi$

The area between them is:

$100\pi - 49\pi = 51\pi$

16. 3 $V = lwh$
$V = 24 \times 12 \times 9 = 2{,}592$ cu. in.

17. 5 The volume of a cylinder is

$V = \pi r^2 h$

$V = \frac{22}{7} \times 7 \times 7 \times 15$

$V = 2{,}310$ cu. in.
$2{,}310 \div 231 = 10$ gallons

18. 1 First convert all dimensions to yards.

$3 \times 4 \times \frac{1}{6} = 2$ cu. yds.

19. 2 $9'' = \frac{3}{4}$ of a foot.

$9' \div \frac{3}{4} = 9 \times \frac{4}{3}$ or 12 tiles wide.

$12' \div \frac{3}{4} = 12 \times \frac{4}{3}$ or 16 tiles wide.

$12 \times 16 = 192$

**20.** 5 The area of the pie is $A = \pi r^2$ or $3.14 \times 4.5 \times 4.5 \doteq 63.6$. 45° is $\frac{1}{8}$ of 360° for the entire circle.

$$\frac{1}{8} \times 63.6 \doteq 8 \text{ sq. in.}$$

**21.** 4 There are 100 cm in a meter. $100 \times 100 = 10{,}000$ sq. cm.

**22.** 4 The volume of the cube is $V = 5 \times 5 \times 5 = 125$ cu. in. The volume of the hole is $V = \pi r^2 h = 3.14 \times 1 \times 1 \times 5 = 15.7 \doteq 16$.

$$125 - 16 = 109$$

**23.** 5 Using the Pythagorean Theorem,

$$x^2 = 90^2 + 90^2$$
$$x^2 = 8{,}100 + 8{,}100$$
$$x^2 = 16{,}200$$
$$x \doteq 127$$

**24.** 3 Use $C = \pi d$

$$C = 3.14 \times 8{,}000$$
$$C = 25{,}120 \text{ or (rounded off) } 25{,}000 \text{ miles}$$

**25.** 2

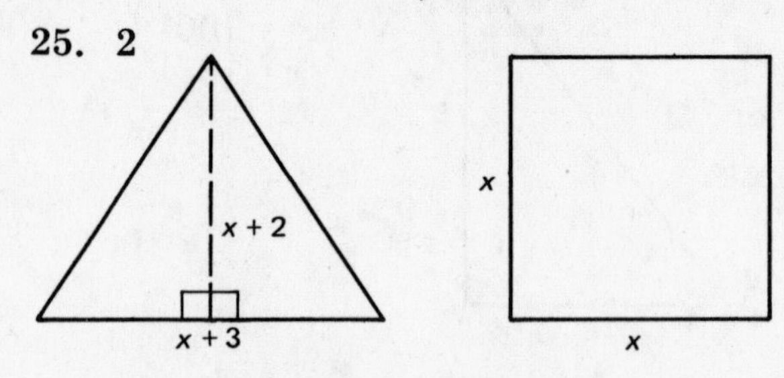


The area of the square is $x^2$.
The area of the triangle is $\frac{1}{2}(x + 2)(x + 3)$.

$$x^2 = \frac{1}{2}(x + 2)(x + 3)$$
$$2x^2 = (x + 2)(x + 3)$$
$$2x^2 = x^2 + 5x + 6$$
$$x^2 - 5x - 6 = 0$$
$$(x - 6)(x + 1) = 0$$
$$x = 6 \quad \text{or} \quad x = -1$$

Since $x = -1$ is not possible, $x = 6$.

## EVALUATION CHART

Multiply the number of correct answers by 100 and divide by the number of questions in the Trial Test to arrive at your score. ____________

| Score | Rating |
|---|---|
| 90–100 | Excellent |
| 80–89 | Very Good |
| 70–79 | Average |
| 60–69 | Passing |
| 59 or Below | Very Poor |

# Chapter 9

# TABLES AND GRAPHS

# 9.1 Reading and Using a Table

Whenever you pick up a newspaper or read a news magazine, you are likely to see a *table* or *graph*. These devices are used to give the reader a fast understanding of groups of numbers. The ability to read and understand a table can help you in many ways. Here are some hints on interpreting and using tables to answer test questions.

1. Read the question carefully. Make sure you understand what information you are looking for.
2. When reading a table, use the proper row and column. Mistakes often happen through misreading.
3. Read carefully the title of the table and all headings to make sure you know what information is given.
4. Before you make any comparisons, make sure you know what units — pounds, feet, hours — are used. Are the units in the table the same as those given in the question?

*Examples:* A table you see every day in the newspaper is the baseball standings. Here are the standings for the Eastern Division of the National League in a recent year:

NATIONAL LEAGUE

East

| | W | L |
|---|---|---|
| Pittsburgh | 61 | 38 |
| New York | 54 | 44 |
| Chicago | 53 | 49 |
| St. Louis | 48 | 50 |
| Montreal | 45 | 52 |
| Philadelphia | 38 | 62 |

($W$ means the number of games won and $L$ means the number of games lost.)

1. What percent of its games did Chicago win?

   (1) 51% (2) 52% (3) 53% (4) 49% (5) 0.52%

   Chicago won 53 games and lost 49. Since it played 102 games, it won $\frac{53}{102}$ of the games played. Divide 102 into 53: the answer is 0.520. To convert to a percent, move the decimal point two places to the right. The correct answer is choice (2) or 52%.

2. In an entire season, 162 games are played. How many games remain for Montreal?

   (1) 117 (2) 110 (3) 65 (4) 75 (5) 55

   Montreal has played 45 + 52 or 97 games. 162 - 97 = 65 games remaining to be played. The answer is choice (3).

3. If St. Louis wins 45 of its remaining games, what percent of all its games will it have won?

   (1) 49% (2) 0.49% (3) 0.57% (4) 57% (5) 61%

   If St. Louis wins 45 more games, it will have won 45 + 48 or 93 games out of 162. Converting $\frac{93}{162}$ to a percent, we obtain choice (4) or 57%. Notice that choice (3), 0.57%, is not correct. It's easy to forget to move the decimal point two places to the right.

The stock market column is another table that appears in most newspapers. Below is a list of a few stocks from a market report.

| | Sales (hds) | High | Low | Close | Net Change |
|---|---|---|---|---|---|
| Am T T | 1444 | $42\frac{7}{8}$ | $42\frac{5}{8}$ | $42\frac{5}{8}$ | $-\frac{1}{4}$ |
| Control Data | 388 | $62\frac{1}{2}$ | $61\frac{3}{4}$ | $61\frac{3}{4}$ | $-1\frac{1}{4}$ |
| IBM | 272 | $389\frac{3}{4}$ | $382\frac{3}{4}$ | $388\frac{3}{4}$ | $+4\frac{1}{4}$ |
| RCA | 292 | $36\frac{3}{4}$ | $36\frac{1}{4}$ | $36\frac{5}{8}$ | — |
| Xerox | 230 | $137\frac{3}{4}$ | $135\frac{1}{4}$ | $137\frac{3}{4}$ | $+2\frac{3}{4}$ |

To understand the headings, read from left to right. *Sales (hds)* means how many shares of stock were sold in hundreds. For example, 272 hundreds or 27,200 shares of IBM stock were sold. The next three numbers represent the dollar costs of one share of stock. Thus $36\frac{3}{4}$ means $36.75 for one share of stock. *High* means the highest price of the day. *Low* is the lowest price of the day, and *Close* means what the stock sold for at the close of the day. *Net Change* means how much the value of the stock changed from the previous day.

1. What was the lowest price for one share of Xerox stock?

   (1) $135.25 (2) $2.75 (3) $230 (4) $137.75 (5) $36.63

   Read across the Xerox line to the *Low* price, $135\frac{1}{4}$. This means $135.25 for one share or choice (1).

2. Which stock had the greatest change in the value of each share?

   (1) Am T T (2) Control Data (3) IBM (4) RCA (5) Xerox

   Choice (3). IBM changed + $4\frac{1}{4}$ or went up $4.25 for each share.

3. Which stock had the greatest *percent* of change for the day?

   (1) Am T T (2) Control Data (3) IBM (4) RCA (5) Xerox

   This is hard to answer, since each change must be compared to the closing price. Control Data (choice (2)) had the greatest change compared to its closing price (2.02%).

4. What is the approximate total value of all the Xerox stock sold that day?

   (1) $31,280 (2) $3,128,000 (3) $137 (4) $13,700 (5) $2.75

   230 hundred or 23,000 shares of Xerox were sold. The average price was about $136. $23{,}000 \times 136 \doteq 3{,}128{,}000$ or choice (2).

# 9.2 Bar Graphs

Bar graphs are usually used to compare measurements. Often, the numbers are large, and the graph gives a quick picture that's easy to understand.

*Example:* On page 157 is a bar graph which shows our trade with Russia during a ten-year period.

1. In what year did we import the most from Russia?

   (1) 1971 (2) 1968 (3) 1969 (4) 1970 (5) 1964

   The gray bar shows imports. The largest gray bar was for 1970 — choice (4).

2. What was the difference between U.S. exports and imports in 1961?

   (1) $9,000,000 (2) $19,000,000 (3) $39,000,000 (4) $24,000,000 (5) $104,000,000

   Choice (2). Exports were about $43,000,000, while imports were about $24,000,000. The difference is $19,000,000.

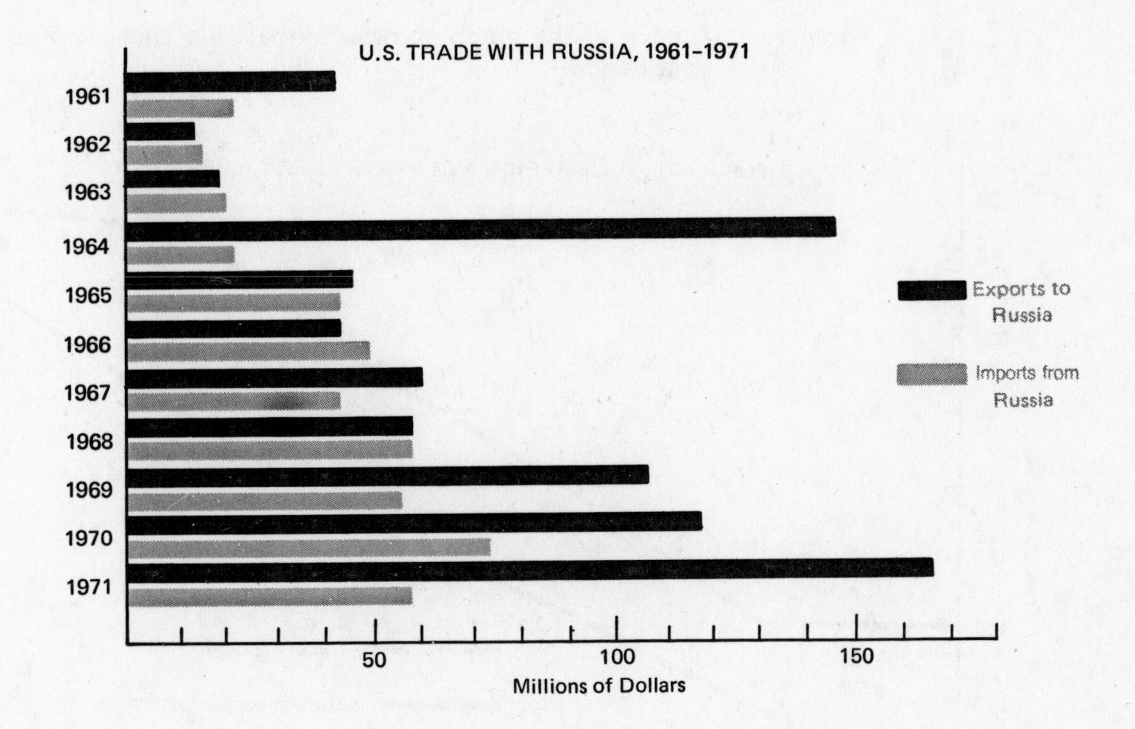


3. Approximately how many millions of dollars worth of goods were imported from Russia in 1961 through 1964?

(1) 82 (2) 23 (3) 22 (4) 100 (5) 150

Choice (1). Imports during these years, shown on the gray bars, are 23, 17, 20, and 22 million dollars. The sum is $82,000,000.

4. What is the difference between the highest and lowest imports from Russia over the 10-year period?

(1) $17,000,000 (2) $73,000,000 (3) $53,000,000 (4) $37,000,000 (5) $56,000,000

Choice (5). Largest imports were in 1970, $73,000,000. Smallest were in 1962, about $17,000,000. $73,000,000 - $17,000,000 = $56,000,000.

# 9.3 Line Graphs

Line graphs are most frequently used to show a change in a measurement. Temperature, weight and height of individuals, test scores, and changes in the stock market are often illustrated with line graphs.

*Example*: Here is a line graph showing population changes over a 70-year period.

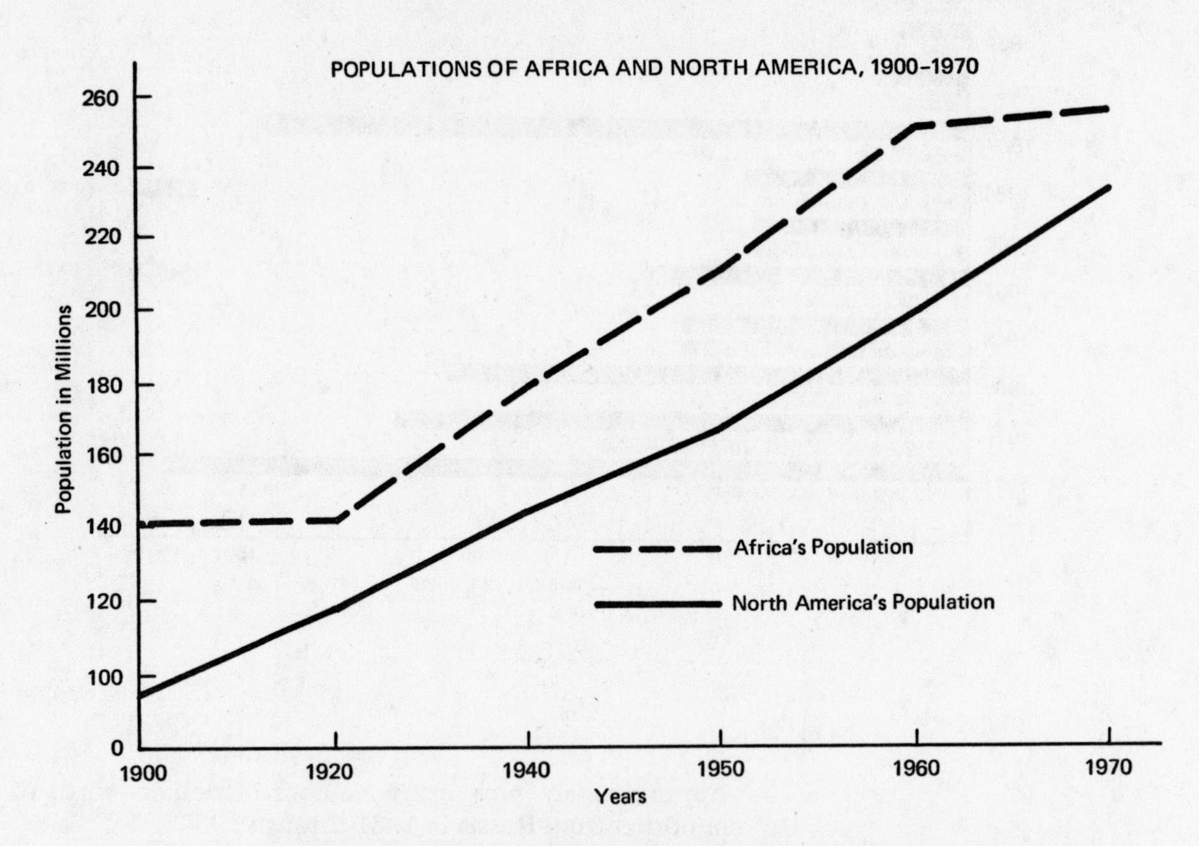


1. During which year was the difference between Africa's population and North America's population greatest?

   (1) 1900 (2) 1920 (3) 1940 (4) 1960 (5) 1970

   Choice (4). The dotted line and the solid line are farthest apart in 1960. They are almost—but not quite— as far apart in 1900.

2. Estimate the North American population in 1945.

   (1) 146 million (2) 165 million (3) 155 million (4) 178 million (5) 205 million

   Choice (3). Draw a vertical line midway between 1940 and 1950. Where it hits the solid line, read the population at the left. This is less than 160 million. The best estimate is 155 million.

3. What was the difference in the population of North America and Africa in 1950?

   (1) 206 million (2) 167 million (3) 29 million (4) 39 million (5) 49 million

   Choice (4). Africa's population was about 206,000,000 while North America's was about 167 million. The difference is 39 million.

4. What was the combined population of North America and Africa in 1950?

(1) 473 million (2) 167 million (3) 273 million (4) 206 million (5) 373 million

Choice (5). North America's population was 167 million and Africa's population was 206 million. The total is 373 million.

5. What was the percent of increase in Africa's population from 1900 to 1960?

(1) 80% (2) 8% (3) 4% (4) 44% (5) 60%

Choice (1). In 1900, Africa's population was 141 million, while in 1960 it was 254 million. To find the percent of increase, compare the actual increase (113 million) to the original population (141 million). $\frac{113}{141} = 0.801$ or about 80%.

# 9.4 Circle Graphs

Circle graphs are ordinarily used for showing how a quantity is divided. Your tax dollar or your monthy budget could be pictured on a circle graph.

*Example*: Here is a breakdown of U.S. population, represented on a circle graph.

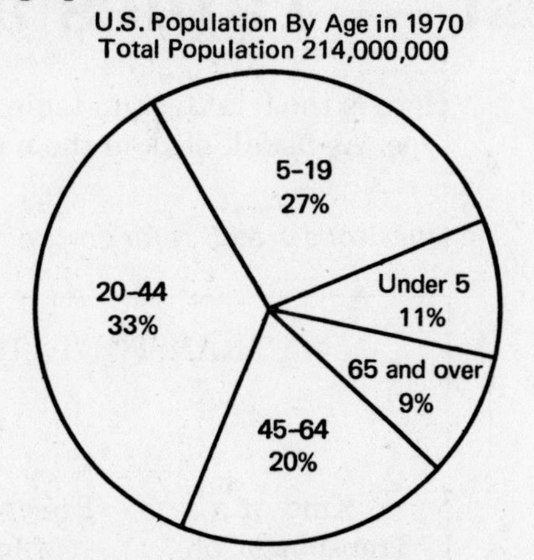


1. Which age group is the largest in the population?

(1) under 5 (2) 5 to 19 (3) 20 to 44 (4) 45 to 64 (5) 65 and over

Choice (3). The 20 to 44 age group contains 33% of the population.

2. What percent of the population is under 65 years of age?

(1) 20% (2) 60% (3) 91% (4) 80% (5) 71%

Choice (3). Since 9% are 65 or over, 91% are under 65. Also, you may add the percents in the other groups:

$$11\% + 27\% + 33\% + 20\% = 91\%.$$

3. Approximately how many people are in the age group 45 to 64?
(1) 20 (2) 42.8 million (3) 33 (4) 71.3 million (5) 107 million

Choice (2). In this age group there are 20% or $\frac{1}{5}$ of the population. $\frac{1}{5} \times 214{,}000{,}000$ is about 42.8 million.

4. What fractional part of the population is in the 20 to 44 age group?
(1) $\frac{1}{3}$ (2) $\frac{1}{4}$ (3) $\frac{1}{5}$ (4) $\frac{1}{33}$ (5) $\frac{1}{10}$

Choice (1), since 33% belong to this group. 33% means $\frac{33}{100}$ or about $\frac{1}{3}$.

5. About how many more people are in the 20 to 44 age group than in the 5 to 19 age group?
(1) about 71 million (2) 6 (3) about 13 million (4) 60 (5) about 54 million

Choice (3). 33% is 6% more than 27%. 6% of 214 million is the same as $0.06 \times 214{,}000{,}000$. This is 12,840,000 or close to 13 million. Be careful to avoid the trap of answer (2). Answer (2) is 6, which is the *percent* difference between the two groups.

# Trial Test — Tables and Graphs

Here is the trial test on tables and graphs. There is a set of examples for each type. As usual, darken the box beneath the number of the correct answer.

*Questions 1 to 5 refer to the table below.*

TRANSPORTATION: ACCIDENT DEATH RATES

For a Recent Year

| Kind of Transportation | Passenger Miles | Passenger Deaths | Death Rate (per 100,000 miles) |
|---|---|---|---|
| Passenger cars | 1,190,000,000 | 26,800 | 2.3 |
| Passenger cars on turnpikes | 28,000,000 | 330 | 1.2 |
| Buses | 55,700,000 | 90 | 0.16 |
| Railroads | 20,290,000 | 20 | 0.10 |
| Air transport | 35,290,000 | 123 | 0.35 |

1. What form of transportation appears to be least hazardous?
(1) Passenger cars (3) Buses
(2) Passenger cars on turnpikes (4) Railroads
(5) Air transport

1. 1 || 2 || 3 || 4 || 5 ||

2. Which form of transportation resulted in the greatest number of deaths?
(1) Passenger cars (3) Buses
(2) Passenger cars on turnpikes (4) Railroads
(5) Air transport

2. 1 || 2 || 3 || 4 || 5 ||

3. About how many times as great is the death rate for passenger cars as for railroads?
(1) 23 (4) 120
(2) 2.3 (5) 12
(3) 230

3. 1 || 2 || 3 || 4 || 5 ||

4. The number of passenger miles by passenger cars is approximately how many times that of railroads?
(1) 6 (4) 6,000
(2) 60 (5) 1,340
(3) 600

4. 1 || 2 || 3 || 4 || 5 ||

5. The total number of deaths in forms of transportation other than passenger cars was
(1) 0.61 (4) 563
(2) 11,280,000 (5) 27,333
(3) 233

5. 1 || 2 || 3 || 4 || 5 ||

*Question 6 to 10 refer to the bar graph below.*

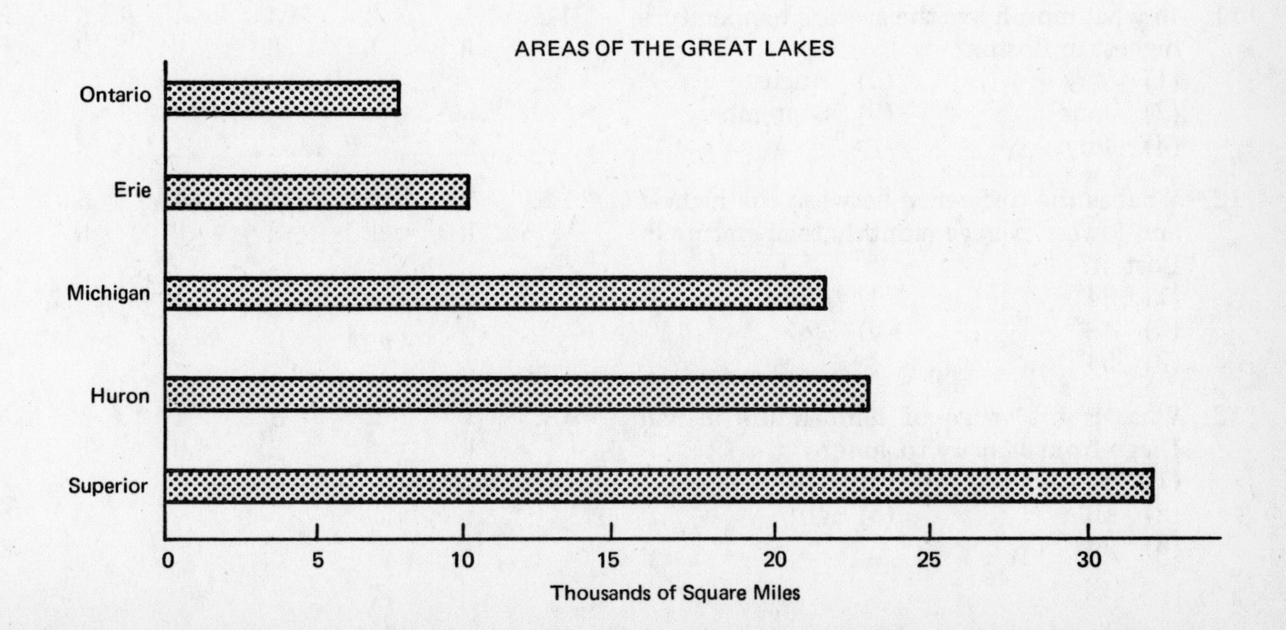

6. About how much larger is Lake Superior than Lake Erie?
(1) 32,000 sq. mi. (4) 15,000 sq. mi.
(2) 22,000 sq. mi. (5) 14,000 sq. mi.
(3) 10,000 sq. mi.

6. 1 || 2 || 3 || 4 || 5 ||

7. Lake Michigan is about how many times as large as Lake Ontario?
(1) 3 (4) 15,000
(2) 30 (5) 15
(3) 300

7. 1 || 2 || 3 || 4 || 5 ||

8. What is the total area of the Great Lakes?
(1) 95 sq. mi. (4) 32,000 sq. mi.
(2) 95,000 sq. mi. (5) 320,000 sq. mi.
(3) 950,000 sq. mi.

8. 1 || 2 || 3 || 4 || 5 ||

9. What is the ratio of the area of Lake Erie to that of Lake Superior
(1) $\frac{5}{16}$ (4) $\frac{1}{5}$
(2) $\frac{16}{5}$ (5) $\frac{1}{32}$
(3) $\frac{1}{4}$

9. 1 || 2 || 3 || 4 || 5 ||

10. Which lake is $2\frac{1}{3}$ times as large as Lake Erie?
(1) Lake Ontario (4) Lake Superior
(2) Lake Michigan (5) None of these
(3) Lake Huron

10. 1 || 2 || 3 || 4 || 5 ||

*Questions 11 to 15 refer to the line graph on page 163.*

11. In what month was the average temperature highest in Boston?
(1) May (4) August
(2) June (5) September
(3) July

11. 1 || 2 || 3 || 4 || 5 ||

12. What is the difference between the highest and lowest average monthly temperature in Boston?
(1) 34° (4) 72°
(2) 48° (5) 28°
(3) 54°

12. 1 || 2 || 3 || 4 || 5 ||

13. What is the range of temperature in San Diego from January to June?
(1) 5° (4) 30°
(2) 10° (5) 40°
(3) 20°

13. 1 || 2 || 3 || 4 || 5 ||

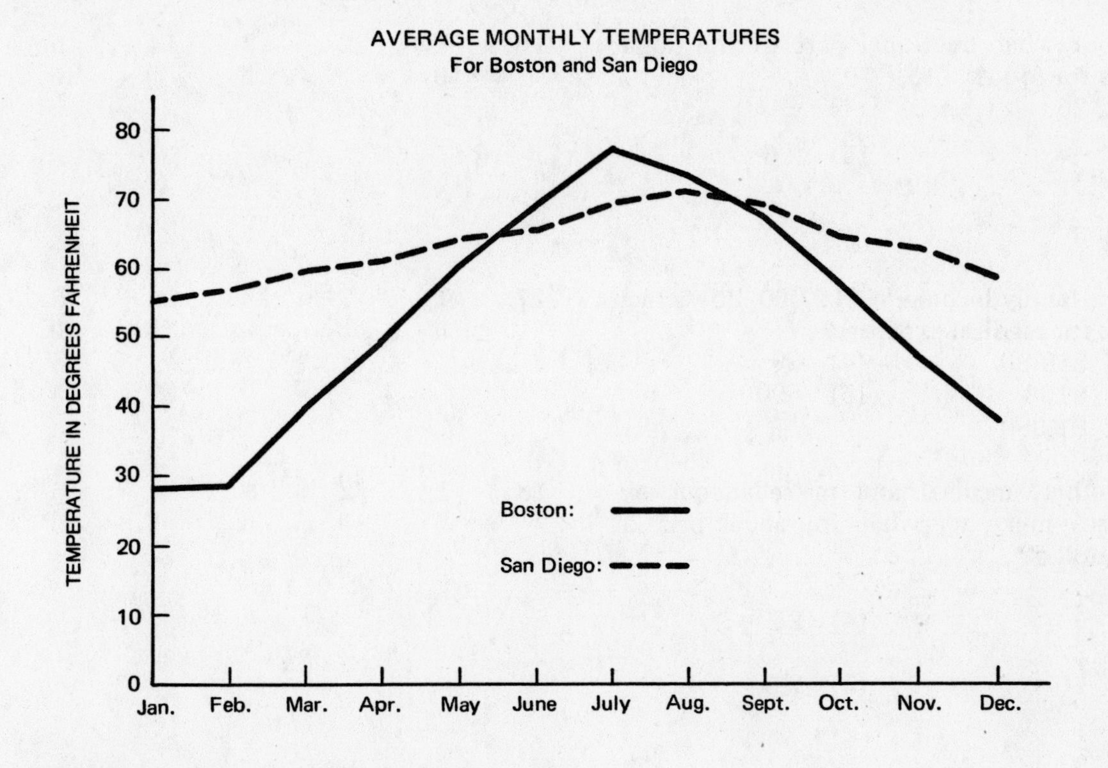


**14.** What is the difference between the lowest average monthly temperatures in the two cities?

(1) 20°
(2) 37°
(3) 17°
(4) 7°
(5) 27°

14. 1 || 2 || 3 || 4 || 5 ||

**15.** The lowest temperature in either of the two cities during the year was

(1) 28°
(2) 55°
(3) 34°
(4) 57°
(5) Cannot be determined from the information given.

15. 1 || 2 || 3 || 4 || 5 ||

*For questions 16 to 20, refer to the circle graph.*

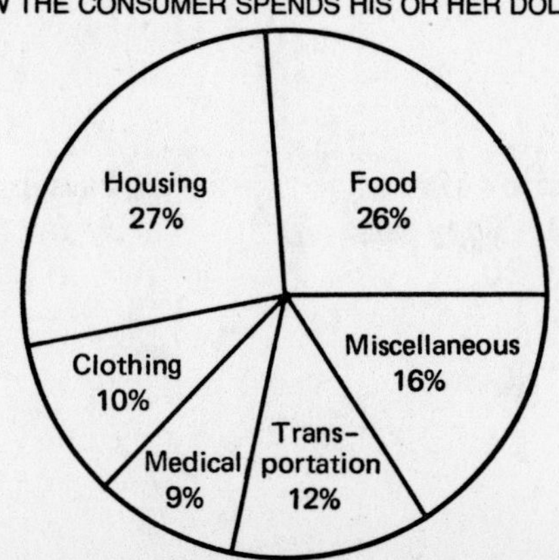

16. About what fractional part of the dollar goes for food?
(1) 26
(2) $\frac{1}{4}$
(3) $\frac{1}{5}$
(4) $\frac{1}{3}$
(5) 2.6

16. 1 || 2 || 3 || 4 || 5 ||

17. On a family income of $12,000, how much goes for medical expenses?
(1) $10.80
(2) $108
(3) $1,080
(4) $9
(5) $900

17. 1 || 2 || 3 || 4 || 5 ||

18. Together, medical and miscellaneous expenses make up what fractional part of the dollar?
(1) $\frac{1}{4}$
(2) $\frac{1}{5}$
(3) $\frac{1}{3}$
(4) 25
(5) 2.5

18. 1 || 2 || 3 || 4 || 5 ||

19. On an income of $12,000, how much will the average family spend on food and housing?
(1) $63.60
(2) $636
(3) $6,360
(4) $53
(5) $5,300

19. 1 || 2 || 3 || 4 || 5 ||

20. With a family income of $14,000, how much more is spent on housing than on transportation?
(1) $15
(2) $21
(3) $210
(4) $2,100
(5) $21,000

20. 1 || 2 || 3 || 4 || 5 ||

---

**SOLUTIONS**

1. 4 Railroads, with a death rate of only 0.10, appear to be the safest. They also had the fewest deaths with only 20.

2. 1 Passenger cars had 26,800 deaths. This is by far the largest number.

3. 1 The death rate for passenger cars is 2.3 and for railroads it is 0.10.

$$\frac{2.3}{0.10} = 23.$$

4. 2 Comparing the number of passenger car miles to that of railroads is most easily done by estimation.

$$1{,}190{,}000{,}000 \doteq 1{,}200{,}000{,}000$$
$$20{,}290{,}000 \doteq 20{,}000{,}000$$

Comparing $\frac{1{,}200{,}000{,}000}{20{,}000{,}000} \doteq 60$

5. 4 563 is obtained by adding all the deaths except for passenger cars.

$$330 + 90 + 20 + 123 = 563$$

6. 2 Lake Superior has about 32,000 sq. mi. Lake Erie has about 10,000 sq. mi. The difference is 22,000 sq. mi.

7. 1 Lake Michigan has about 22,500 sq. mi. Lake Ontario has about 7,500 sq. mi. The ratio $\frac{22{,}500}{7{,}500}$ is the same as 3.

8. 2 The lakes contain in order 7,500, 10,000, 22,500, 23,000, and 32,000 sq. mi. The sum of these is 95,000 sq. mi.

9. 1 The two areas are 10,000 and 32,000 sq. mi. $\frac{10{,}000}{32{,}000}$ is the same as $\frac{5}{16}$.

10. 3 Lake Huron is the best answer. Its area of 23,000 sq. mi. is about $2\frac{1}{3}$ times that of Lake Erie.

$$2\frac{1}{3} \times 10{,}000 \doteq 23{,}000.$$

11. 3 July was the hottest at 76°.

12. 2 July's temperature was 76° while January and February averaged 28°. The difference is 48°.

13. 2 Range means the difference from highest to lowest. The temperatures in San Diego varied from 55° to 65°, a range of 10°.

14. 5 The lowest temperature for San Diego was 55° and for Boston 28°. The difference is 27°.

15. 5 This answer cannot be determined. The graph deals with average monthly temperatures, not the *daily* highs and lows.

16. 2 Food accounted for 26% of the money spent. 26% means $\frac{26}{100}$, which is close to $\frac{1}{4}$ or $\frac{25}{100}$.

17. 3 Medical expenses were 9% of the total expenditures. 9% of $12,000 means 0.09 × 12,000 or $1,080. Don't forget the position of the decimal point.

18. 1 Medical and miscellaneous expenses were 16% and 9%.

$$16\% + 9\% = 25\% \text{ or } \frac{1}{4}$$

19. 3 Housing and food together make up 53% of the expenditures. 53% of $12,000 is $6,360.

20. 4 Housing used 27% while transportation accounted for 12%. The difference is 15%. 15% of $14,000 is $2,100.

**EVALUATION CHART**

Multiply the number of correct answers by 100 and divide by the number of questions in the Trial Test to arrive at your score. ____________

| Score | Rating |
|---|---|
| 90-100 | Excellent |
| 80-89 | Very Good |
| 70-79 | Average |
| 60-69 | Passing |
| 59 or Below | Very Poor |

# Chapter 10

# WORD PROBLEMS

# Reading a Word Problem

The whole point of study in mathematics is to be able to solve problems and to handle real-life situations.

Throughout this book you have already attacked a variety of problems in arithmetic, algebra, and geometry. Think of this next set as a warm-up for the sample tests in Chapter 11. The following problems represent a mixture of the areas covered by these tests.

The secret of success in problem solving is *reading*. In your attack, each problem should actually be read three times with a different purpose for each.

1. A fast first reading to see what the problem is.
2. A slow second reading, with emphasis on key words, usually to set up an equation or mathematical sentence to solve the problem.
3. A reading done after the problem has been solved, either to check the answer if possible or to make sure the answer is reasonable.

There are times when it is not necessary to actually solve the problem in a multiple-choice question; you can merely try out each answer in turn to see if it fits. While random guessing is not recommended, if you can reject, say, two of five possibilities, it is a good idea to make your best guess. Wherever possible, draw a picture to illustrate the problem.

*Example* 1: If 8 quarts of antifreeze are necessary to protect a 12-quart system of a car to $-34°$, how much antifreeze would be needed to protect a small car with a 5-quart capacity to $-34°$?

(1) 6 quarts

(2) 5 quarts

(3) $2\frac{3}{4}$ quarts

(4) $3\frac{1}{3}$ quarts

(5) 4 quarts

*Solution*: Draw a picture.

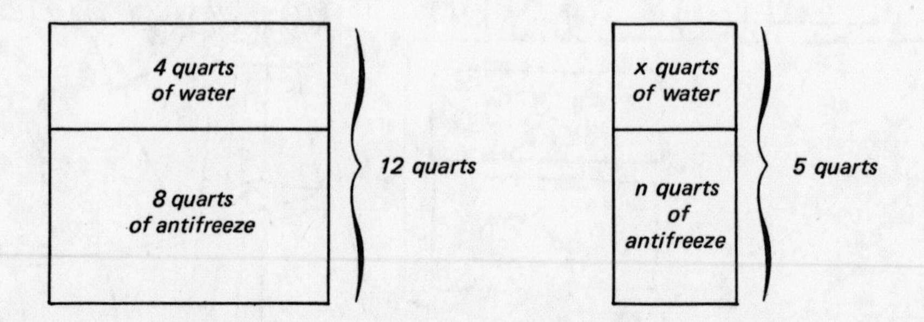

Clearly, a proportion will solve the problem.

$$\frac{8}{12} = \frac{n}{5}$$

$$12n = 40$$

$$n = \frac{40}{12}$$

$n = 3\frac{1}{3}$ quarts, or choice (4).

This is the straightforward approach. But if you were in trouble in the problem, at least you could look at the answers and discard choices 1 and 2 as illogical. Then it would be right to make a guess.

> To summarize:
> 1. Read the problem three times (with different aims).
> 2. Use the answers as clues.
> 3. Draw pictures where possible.

*Example* 2: A room is 19 feet long, 10 feet wide, and 8 feet high. If you wanted to paint the walls and ceiling, how many square feet would you need to paint?

(1) 238
(2) 304
(3) 464
(4) 654
(5) 794

*Solution*: First, draw a picture.

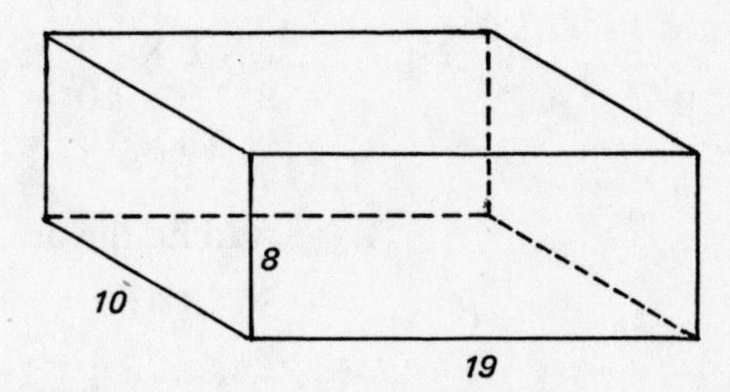


You might even want to draw a picture for each different surface you want to paint.

2 walls

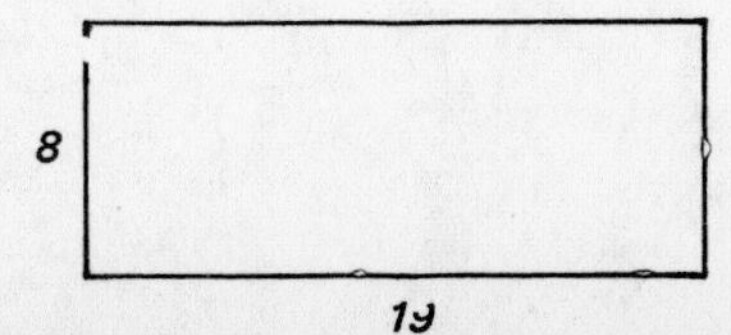

2 other walls

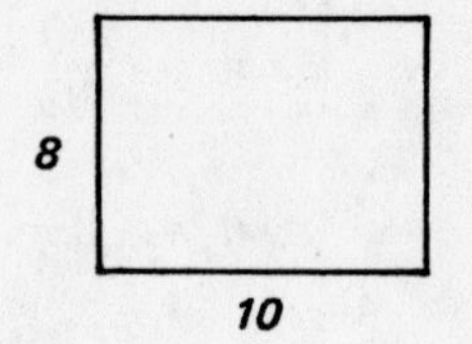


ceiling

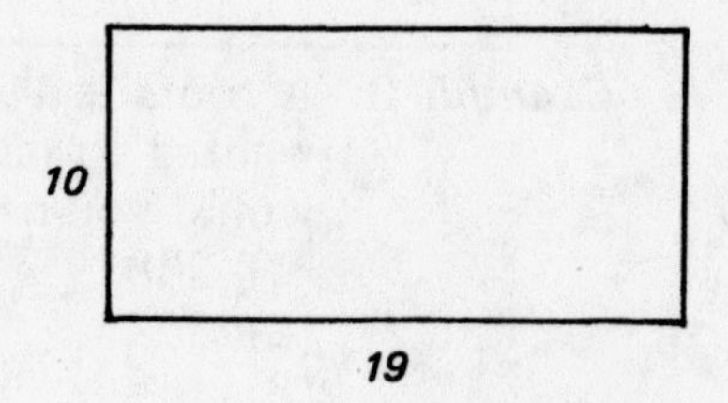


Then, find the area of each:

$2 \times 8 \times 19 = 304$
$2 \times 8 \times 10 = 160$
$10 \times 19 = 190$

Adding the areas,

$304 + 160 + 190 = 654$, or choice (4).

# Trial Test — Word Problems

**1.** A tennis racket priced at $30 is on sale at 10% off the list price. If there is a 5% sales tax in this area, how much will the racket cost?
(1) $27
(2) $28.35
(3) $31.50
(4) $29.35
(5) $28

1. 1 2 3 4 5

**2.** Ginger ale is on sale at two for $0.79. How much change would you receive from a $10 bill if you bought a dozen bottles?
(1) $9.21
(2) $9.31
(3) $4.74
(4) $6.26
(5) $5.26

2. 1 2 3 4 5

**3.** Twenty, divided by one-half, plus ten is equal to
(1) 20
(2) 30
(3) 40
(4) 50
(5) 25

3. 1 2 3 4 5

**4.** A cube, 3″ on each edge, is painted entirely red. It is then cut apart into 1″ cubes. How many of these will have paint on exactly two faces?
(1) none
(2) 8
(3) 4
(4) 12
(5) 6

4. 1 2 3 4 5

**5.** A rectangular block of wood, 3″ by 4″ by 5″, is cut from a cubic foot of wood weighing 72 pounds. How much does the small block weigh?
(1) 2½ pounds
(2) 5 pounds
(3) 24 pounds
(4) 36 pounds
(5) 7½ pounds

1 2 3 4 5

**6.** When the time is exactly 6:30 P.M., what is the measure of the angle between the hands of the clock?
(1) 0°
(2) 15°
(3) 30°
(4) 24°
(5) 36°

6. 1 2 3 4 5

7. One factor of $2x^2 + 7x - 15$ is $2x - 3$. What is the other factor?
(1) $2x - 5$
(2) $x + 5$
(3) $x - 5$
(4) $2x + 3$
(5) $2x + 5$

7. 1 2 3 4 5

8. John budgeted half his month's salary for mortgage, $\frac{1}{4}$ for food, $\frac{1}{6}$ for clothing, and put the rest ($80) into savings. What was his monthly salary?
(1) $960
(2) $160
(3) $320
(4) $480
(5) $720

8. 1 2 3 4 5

9. A right triangle has one leg 8″ long and the hypotenuse is 10″ long. What is the length of the side of a square with the same area as the triangle?
(1) $\sqrt{80}''$
(2) $2\sqrt{6}''$
(3) 9″
(4) $4\sqrt{3}''$
(5) 6″

9. 1 2 3 4 5

10. A man took a trip by car, starting at 11 A.M. at 50 mph. He arrived at his destination at 2 P.M. A bicyclist started from the same place at the same time at an average speed of 15 mph. At what time will he reach the same destination?
(1) 6 P.M.
(2) 7 P.M.
(3) 8 P.M.
(4) 9 P.M.
(5) 10 P.M.

10. 1 2 3 4 5

11. My average grade on a set of 5 tests was 88%. I can only remember that the first 4 grades were 78%, 86%, 96%, and 94%. What was my fifth grade?
(1) 82%
(2) 84%
(3) 86%
(4) 88%
(5) 90%

11. 1 2 3 4 5

12. Each edge of a 5″ cube is increased by 20%. What is volume of the new cube?
(1) 6 cu. in.
(2) 36 cu. in.
(3) 150 cu. in.
(4) 216 cu. in.
(5) 476 cu. in.

12. 1 2 3 4 5

13. When 39 is raised to the 100th power, the unit's digit of the answer will be
(1) 1
(2) 9
(3) 7
(4) 6
(5) 0

13. 1 2 3 4 5

**14.** In a right triangle, the lengths of the sides are consecutive integers. What is the length of the shortest side?
(1) 1
(2) 2
(3) 3
(4) 4
(5) 5

**14.** 1 || 2 || 3 || 4 || 5 ||

**15.** The average television viewer watches 2.6 hours of TV per day. How many minutes will he watch in a week?
(1) 18.2
(2) 156
(3) 546
(4) 2,184
(5) 1,092

**15.** 1 || 2 || 3 || 4 || 5 ||

**16.** Katie has a $10 roll of nickels, a $10 roll of dimes, and a $10 roll of quarters. How many coins does she have?
(1) 340
(2) 30
(3) 600
(4) 300
(5) 120

**16.** 1 || 2 || 3 || 4 || 5 ||

**17.** Ten ounces of a liquid contain 20% alcohol and 80% water. It is diluted by adding 40 ounces of water. The percent of alcohol in the new solution is
(1) 20%
(2) 2%
(3) 40%
(4) 4%
(5) 10%

**17.** 1 || 2 || 3 || 4 || 5 ||

**18.** Most beer contains 4% alcohol by volume. How many ounces of alcohol are contained in one quart of beer? (One quart = 32 ounces)
(1) 1.28
(2) 12.8
(3) .64
(4) 4
(5) 6.4

**18.** 1 || 2 || 3 || 4 || 5 ||

**19.** A slow-burning candle burns down at a rate of 0.6 inch per hour. How many hours will a 9″ candle last?
(1) 54
(2) 15
(3) 6
(4) 5.4
(5) 150

**19.** 1 || 2 || 3 || 4 || 5 ||

**20.** Two new cars, A and B, have the same list price of $5,000. Car A advertises an 8% discount and a $300 rebate. Car B advertises a 15% discount. Which is the better buy, and by how much?
(1) Car A by $100
(2) Car B by $100
(3) Car A by $50
(4) Car B by $50
(5) They are equal buys.

**20.** 1 || 2 || 3 || 4 || 5 ||

21. A 20″ TV set means a diagonal of 20″. If one edge of its tube is 12″ long, how many square inches of tube area are there?
(1) 144
(2) 192
(3) 256
(4) 120
(5) 240

21. 1 || 2 || 3 || 4 || 5 ||

22. If the number 100 is first reduced by 10%, and the resulting number is increased by 10%, what is the final number?
(1) 99
(2) 100
(3) 98
(4) 101
(5) 110

22. 1 || 2 || 3 || 4 || 5 ||

23. The smallest prime number greater than 200 is
(1) 201
(2) 203
(3) 205
(4) 207
(5) 211

23. 1 || 2 || 3 || 4 || 5 ||

24. While a building casts a shadow 24′ long, a vertical yardstick casts a shadow 2′ long. What is the height of the building?
(1) 36′
(2) 16′
(3) 20′
(4) 24′
(5) 48′

24. 1 || 2 || 3 || 4 || 5 ||

25. The perimeter of a rectangle is 28″. If its length is 8″, how long is a diagonal?
(1) 8″
(2) 6″
(3) 12″
(4) 10″
(5) 14″

25. 1 || 2 || 3 || 4 || 5 ||

## SOLUTIONS

1. 2 10% of \$30 is \$3, so the cost of the racket is \$30 - \$3 = \$27. 5% of \$27 is \$1.35, so the total cost is \$27 + \$1.35 or \$28.35.

2. 5 There are 6 twos in a dozen.
$6 \times \$0.79 = \$4.74$.
$\$10 - \$4.74 = \$5.26$ (the change you should receive).

3. 4 Did you choose (1)? If so, then think about "20 divided by ½."
$$20 \div \frac{1}{2} = 20 \times 2 = 40, \text{ and}$$
$$40 + 10 = 50.$$

4. 4

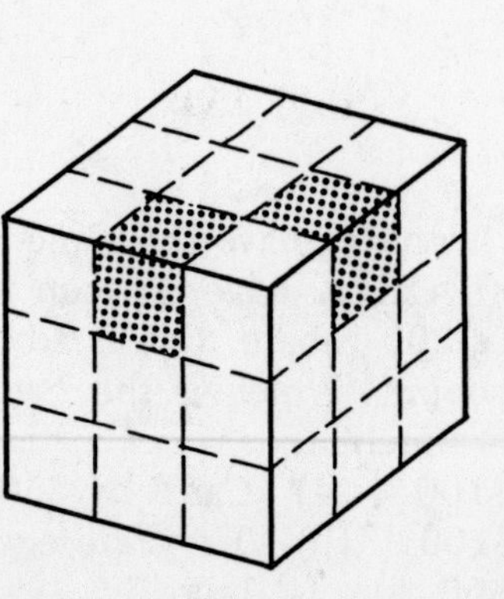

Shown in the picture are two of the small cubes with exactly two faces painted. Carefully counting the total, there are four each on the top and bottom and four more around the sides for a total of 12.

**5.** **1** First convert the inches to feet to obtain $\frac{3}{12} \times \frac{4}{12} \times \frac{5}{12} = \frac{5}{144}$. The volume of the block is $\frac{5}{144}$ cubic feet and $\frac{5}{144} \times 72 = \frac{5}{2}$ or $2\frac{1}{2}$ pounds.

**6.** **2** At 6:30, the hour hand has moved half the way between 6 and 7. This would be $\frac{1}{2}$ of $\frac{1}{12}$ of the circle.

$$\frac{1}{2} \times \frac{1}{12} \times 360 = 15°$$

**7.** **2** $x + 5$ is the other factor.

$$(x + 5)(2x - 3) = 2x^2 - 3x + 10x - 15$$
$$= 2x^2 + 7x - 15$$

**8.** **1**

$$\frac{1}{2} + \frac{1}{4} + \frac{1}{6} = \frac{6}{12} + \frac{3}{12} + \frac{2}{12}$$
$$= \frac{11}{12}$$

This leaves $\frac{1}{12}$ of his income for savings. $\frac{1}{12}$ of his income is $80. Use algebra as follows:

$$\frac{1}{12}(x) = 80$$
$$(12)\frac{1}{12}(x) = 80(12)$$
$$x = \$960 \text{ (the monthly salary)}$$

**9.** **2** Use drawings:

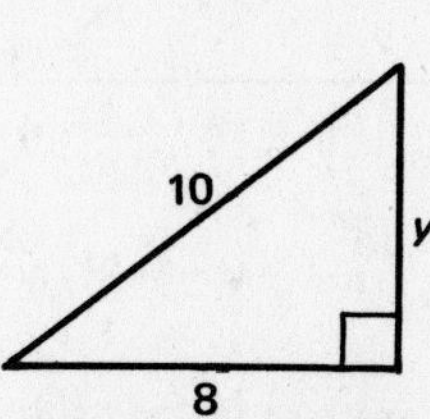


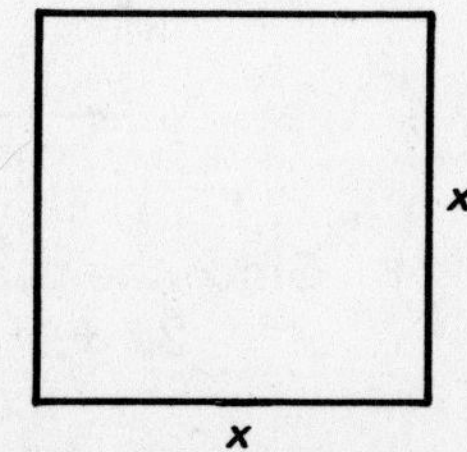


Let the unknown side of the triangle be $y$. $y^2 + 8^2 = 10^2$ or $y^2 + 64 = 100$. Then $y^2 = 36$, so $y = 6$. The area of the triangle is $\frac{1}{2}bh$ or $\frac{1}{2} \times 8 \times 6$, or 24. The area of the square is $x^2$. Thus $x^2 = 24$ and $x = \sqrt{24}$. This can be simplified to $x = \sqrt{4} \times \sqrt{6}$, or $2\sqrt{6}$.

**10.** **4** In three hours, the man traveled $3 \times 50$ or 150 miles. $150 \div 15$ (the bicyclist's rate) = 10 hours. 10 hours after 11 A.M. is 9 P.M.

**11.** **3** The easiest way is again to use algebra, with the unknown test equal to $x$.

$$\frac{78 + 86 + 96 + 94 + x}{5} = 88$$
$$\frac{354 + x}{5} = 88$$
$$354 + x = 440$$
$$x = 86$$

**12.** **4** 20% of 5 is the same as $\frac{1}{5}$ of 5, or 1. The new cube is 6″ on each edge. The volume of a cube is given by $V = e^3$ where $e$ is an edge.

$$6^3 = 6 \times 6 \times 6 = 216.$$

**13.** **1** This is a trick question. You need only concern yourself with the final digit of the number 39.

$$9^2 = 81$$
$$9^3 = 729$$
$$9^4 = \text{---}1$$

From this it's clear that all even powers will end in 1, while all odd powers will end in 9.

**14.** **3** Call the three sides $x$, $x + 1$, and $x + 2$. By the Pythagorean Theorem:

$$x^2 + (x + 1)^2 = (x + 2)^2$$

or

$$x^2 + x^2 + 2x + 1 = x^2 + 4x + 4$$
$$2x^2 + 2x + 1 = x^2 + 4x + 4$$
$$x^2 - 2x - 3 = 0$$
$$(x - 3)(x + 1) = 0$$
$$x = 3, x = -1$$

Clearly it is not possible for $x$ to be $-1$.

**15.** **5** $7 \times 2.6 = 18.2$ *hours* per week.
60 minutes $\times$ 18.2 = 1,092

**16.** **1** A $10 roll of nickels has 200 coins. A $10 roll of dimes has 100 coins. A $10 roll of quarters has 40 coins.

**17.** **4** 20% of 10 ounces is 2 ounces of alcohol. The new mixture has 10 ounces + 40 ounces, or 50 ounces. 2 ounces is 4% of 50 ounces.

$$\frac{2}{50} = \frac{4}{100}$$

**18.** 1 A quart contains 32 ounces. (An easy way to remember this is to think "A pint's a pound the world around." A pint is 16 ounces, and 2 pints equal one quart. Thus a quart is 32 ounces.) 4% of 32 is $0.04 \times 32$ or 1.28 ounces.

**19.** 2 This one is just simple division:

$9 \div 0.6 = 15$

**20.** 4 First look at Car A:

8% of 5,000 is $400
$5,000 - 400 = $4,600
$4,600 - 300 = $4,300 (net price for Car A)

Now Car B:

15% of $5,000 is $750
$5,000 - 750 = $4,250

Car B is the better buy at a savings of $50.

**21.** 2 Draw a diagram to illustrate the problem.

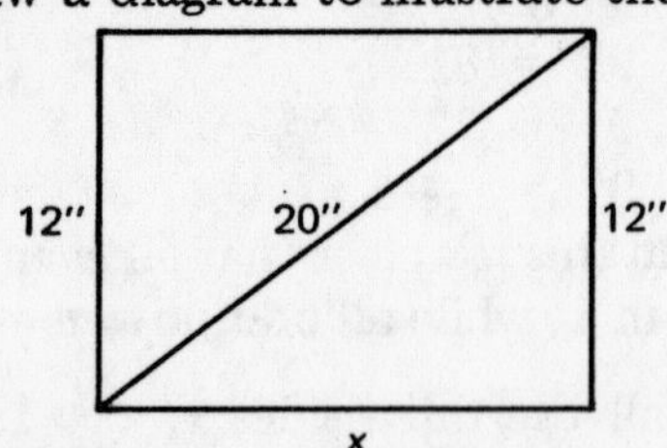


First find the base $x$ by the Pythagorean Theorem.

$$x^2 + 12^2 = 200^2$$
$$x^2 + 144 = 400$$
$$x^2 = 256$$

So $x = 16$

Then $A = l \times w$

$A = 12'' \times 16'' = 192$ sq. in.

**22.** 1 This is another trick question. If you reduce 100 by 10%, you reduce it by 10 to get 90. Now increase 90 by 10% of *90* which is 9.

$90 + 9 = 99$

**23.** 5 201 and 207 are each divisible by 3. 203 is divisible by 7, and 205 is divisible by 5. Only 211 has no divisors other than itself. Thus 211 is prime.

**24.** 1 Use drawings to illustrate the problem. Remember that a yardstick is 3′ long.

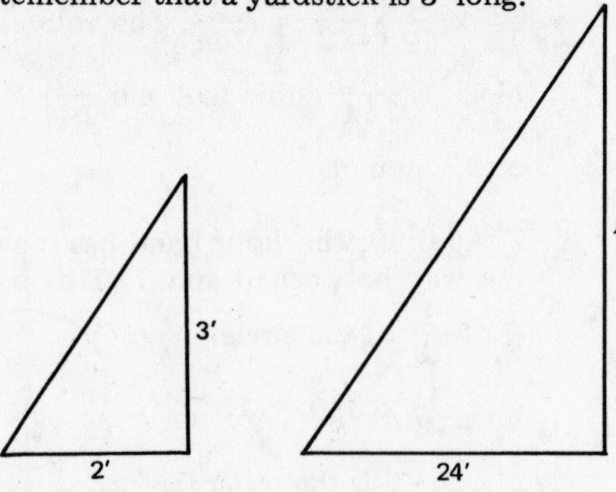


By using the idea that the triangles are similar, and corresponding sides are in proportion:

$$\frac{3}{x} = \frac{2}{24}$$

$$72 = 2x$$
$$x = 36$$

**25.** 4 Again draw a diagram.

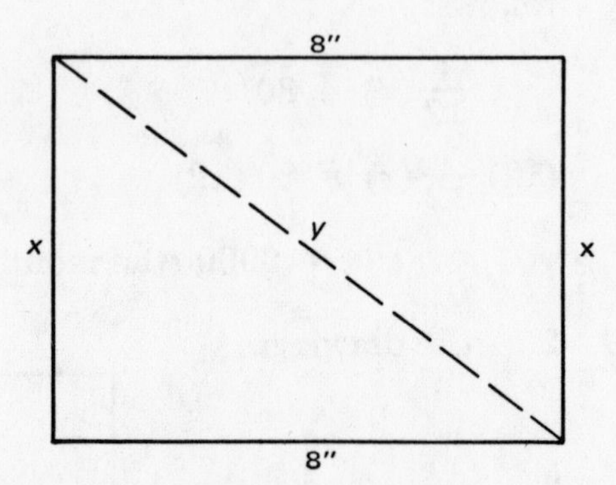


Since the perimeter is 28″,

$$2x + 16 = 28$$
$$2x = 12$$
$x = 6''$ is the width.

Now, use the Pythagorean Theorem to find the diagonal.

$$y^2 = 6^2 + 8^2$$
$$y^2 = 36 + 64 = 100$$
$$y = 10$$

## EVALUATION CHART

Multiply the number of correct answers by 100 and divide by the number of questions in the Trial Test to arrive at your score. ________________

| Score | Rating |
|---|---|
| 90-100 | Excellent |
| 80-89 | Very Good |
| 70-79 | Average |
| 60-69 | Passing |
| 59 or Below | Very Poor |

# Chapter 11

# MODEL EXAMINATIONS AND SOLUTIONS

The first test in this chapter may be used as a Diagnostic Test. After you have taken it and tallied your results, you will have a much clearer idea of those areas in which you are weakest. Review these areas with special care.

You are not expected to get all 50 questions correct. Actually, a score of 32 out of 50 shows that you have attained minimum competence.

Remember the test-taking hints given in this book. It is particularly important not to spend too much time on any single question. When you can eliminate a few of the possible answers, it is a good idea to take an educated guess.

Perhaps the most important lesson you can learn will come from the mistakes you make. These will pinpoint areas where you need more study.

# Diagnostic Test— General Mathematical Ability

**1.** Add: 2.32 + 71.4 + 0.003
(1) 73.75
(2) 94.9
(3) 9.49
(4) 76.72
(5) 73.723

**2.** Subtract: 7,527 – 149
(1) 7,378
(2) 7,478
(3) 7,388
(4) 7,488
(5) 7,487

**3.** Multiply: $\frac{5}{6} \times 1\frac{1}{2}$
(1) $\frac{5}{6}$ (2) $\frac{5}{3}$ (3) $\frac{5}{4}$ (4) $\frac{5}{9}$ (5) $\frac{4}{5}$

**4.** Divide: 92.5 ÷ 3.7
(1) 2.5 (2) 25 (3) 0.25 (4) 250 (5) 342.25

**5.** If a bank pays 6% interest per year on savings accounts, how much interest would an account of $1,200 earn in one year?
(1) $7.20 (2) $72 (3) $720 (4) $7,200 (5) $0.75

**6.** If Mr. Siegel bought items costing $3.75, $2.89, and $7.05, how much change should he receive from a $20 bill?
(1) $13.69
(2) $14.69
(3) $5.31
(4) $6.31
(5) $7.31

| | 1 | 2 | 3 | 4 | 5 |
|---|---|---|---|---|---|
| 1. | ‖ | ‖ | ‖ | ‖ | ‖ |
| 2. | ‖ | ‖ | ‖ | ‖ | ‖ |
| 3. | ‖ | ‖ | ‖ | ‖ | ‖ |
| 4. | ‖ | ‖ | ‖ | ‖ | ‖ |
| 5. | ‖ | ‖ | ‖ | ‖ | ‖ |
| 6. | ‖ | ‖ | ‖ | ‖ | ‖ |

**7.** Jim scored 75, 85, 93, 81, and 91 on 5 tests. What was his average score?
(1) 83 (2) 84 (3) 85 (4) 86 (5) 87

**8.** The fact that $3 + 8 = 8 + 3$ is an illustration of what law?
(1) associative
(2) commutative
(3) distributive
(4) symmetric
(5) transitive

**9.** If $2x + 7 = 25$, then $x = ?$
(1) 18 (2) 32 (3) 8 (4) 9 (5) 16

**10.** Mr. Perez works for $2.80 an hour, plus time and a half for all hours over 40. One week he worked 48 hours. What was his salary?
(1) $134.40
(2) $13.44
(3) $112.00
(4) $145.60
(5) $14.56

**11.** Which of the following fractions is less than $\frac{1}{2}$?
(1) $\frac{5}{6}$ (2) $\frac{3}{5}$ (3) $\frac{3}{10}$ (4) $\frac{2}{3}$ (5) $\frac{5}{9}$

**12.** When a certain operation is performed on two numbers, 7 and 8, a third number, 56, is obtained. Using this operation, we call 56 the:
(1) sum
(2) product
(3) difference
(4) quotient
(5) dividend

| | 1 | 2 | 3 | 4 | 5 |
|---|---|---|---|---|---|
| 7. | ‖ | ‖ | ‖ | ‖ | ‖ |
| 8. | ‖ | ‖ | ‖ | ‖ | ‖ |
| 9. | ‖ | ‖ | ‖ | ‖ | ‖ |
| 10. | ‖ | ‖ | ‖ | ‖ | ‖ |
| 11. | ‖ | ‖ | ‖ | ‖ | ‖ |
| 12. | ‖ | ‖ | ‖ | ‖ | ‖ |

**13.** The population of a city is 328,637. What is the population of this city to the nearest thousand?
(1) 328
(2) 329
(3) 329,000
(4) 328,000
(5) 3,000,000

**14.** One inch is about the same as 2.54 centimeters. Approximately how many centimeters are in a foot?
(1) 20.48 (2) 30.48 (3) 0.21 (4) 26.48 (5) 0.42

**15.** 40% is equal to the fraction $\frac{x}{30}$. What is the value of $x$?

(1) $\frac{2}{5}$ (2) 15 (3) 1,200 (4) 120 (5) 12

**16.** Thirteen tenths written as a decimal fraction is:
(1) 0.013 (2) 0.13 (3) 1.3 (4) 13 (5) 130

**17.**

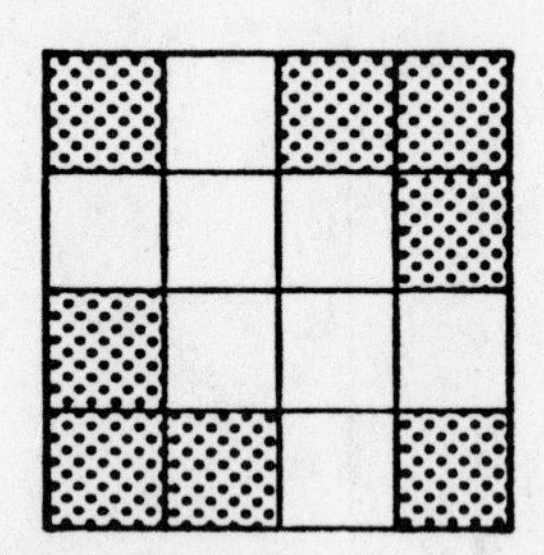

What percent of the squares on the above figure are shaded?

(1) 8 (2) $\frac{1}{2}$ (3) 5 (4) 50 (5) 25

**18.** The cost of a black and white television set recently increased from $160 to $200. What was the percent of increase in the price?

(1) 25 (2) 20 (3) $\frac{1}{4}$ (4) $\frac{1}{5}$ (5) 40

| | 1 | 2 | 3 | 4 | 5 |
|---|---|---|---|---|---|
| 13. | ‖ | ‖ | ‖ | ‖ | ‖ |
| 14. | ‖ | ‖ | ‖ | ‖ | ‖ |
| 15. | ‖ | ‖ | ‖ | ‖ | ‖ |
| 16. | ‖ | ‖ | ‖ | ‖ | ‖ |
| 17. | ‖ | ‖ | ‖ | ‖ | ‖ |
| 18. | ‖ | ‖ | ‖ | ‖ | ‖ |

**19.** A tire sale advertised 20% off on top grade belted tires. If the list price for a particular tire is $40, how much would you have to pay for a set of 4 new tires?
(1) $160
(2) $32
(3) $39.20
(4) $128
(5) $154.80

**20.** What are the prime factors of 24?
(1) $2 \times 3 \times 4$
(2) $4 \times 6$
(3) $2 \times 2 \times 2 \times 3$
(4) $3 \times 8$
(5) $1 \times 2 \times 12$

**21.** If a car traveled 200 miles at an average rate of speed of $r$ miles per hour, the time it took for the trip could be written as:
(1) $\frac{200}{r}$
(2) $\frac{r}{200}$
(3) $200r$
(4) $r$ 200
(5) $200 - 4$

**22.** Two consecutive even numbers have a sum of 30. What is the smaller of the two numbers?
(1) 10 (2) 12 (3) 14 (4) 16 (5) 18

**23.** If $a = 1$, $b = 2$, and $c = 0$, what is the value of $a^3b^2c$?
(1) 0 (2) 4 (3) 12 (4) 7 (5) 5

**24.** The complete factorization of $12x^2 + 4x - 8$ is:
(1) $(3x + 1)(4x - 8)$
(2) $(4x + 4)(3x + 2)$
(3) $4(3x^2 + x - 2)$
(4) $4(3x - 2)(x + 1)$
(5) $4(3x + 2)(x - 1)$

| | 1 | 2 | 3 | 4 | 5 |
|---|---|---|---|---|---|
| 19. | ‖ | ‖ | ‖ | ‖ | ‖ |
| 20. | ‖ | ‖ | ‖ | ‖ | ‖ |
| 21. | ‖ | ‖ | ‖ | ‖ | ‖ |
| 22. | ‖ | ‖ | ‖ | ‖ | ‖ |
| 23. | ‖ | ‖ | ‖ | ‖ | ‖ |
| 24. | ‖ | ‖ | ‖ | ‖ | ‖ |

**25.** The scale on a blueprint reads $\frac{1''}{2}$ to $1'$. If the living room is to be $21'$ long, how long will it be on the blueprint?

(1) 10″ (2) $10\frac{1}{4}''$ (3) $10\frac{1}{2}''$ (4) 11″ (5) 21″

**26.** Which of the following numbers belongs in the solution for $5x > 2x + 7$?

(1) −1 (2) 0 (3) 1 (4) 2 (5) 3

**27.** If a man wishes to cut 5 lengths of board each 2′3″ long, what is the total length of board he must buy?

(1) 10′ (2) 10′3″ (3) 11′ (4) 11′3″ (5) 12′

**28.** How many $\frac{1}{2}$ pint containers can be filled from a 5 gallon can?

(1) 4 (2) 16 (3) 20 (4) 40 (5) 80

**29.** Which of the following numbers is different from the rest?

(1) 0.25 (2) $\frac{1}{4}$ (3) 25% (4) $\frac{1}{4}\%$ (5) $\frac{16}{64}$

**30.** An automobile trip started at 8:35 A.M. and was completed at 2:15 P.M. on the same day. How long did the trip take?

(1) 4 hrs. 20 min.
(2) 4 hrs. 40 min.
(3) 6 hrs. 20 min.
(4) 5 hrs. 40 min.
(5) 5 hrs. 20 min.

**31.** The area of a room is 216 square feet. How many square yards of carpeting will be needed to cover the floor?

(1) 72 (2) 24 (3) 648 (4) 1,944 (5) 27

**32.** What is an angle containing 120° called?

(1) acute
(2) obtuse
(3) right
(4) straight
(5) reflex

| | 1 | 2 | 3 | 4 | 5 |
|---|---|---|---|---|---|
| 25. | ‖ | ‖ | ‖ | ‖ | ‖ |
| 26. | ‖ | ‖ | ‖ | ‖ | ‖ |
| 27. | ‖ | ‖ | ‖ | ‖ | ‖ |
| 28. | ‖ | ‖ | ‖ | ‖ | ‖ |
| 29. | ‖ | ‖ | ‖ | ‖ | ‖ |
| 30. | ‖ | ‖ | ‖ | ‖ | ‖ |
| 31. | ‖ | ‖ | ‖ | ‖ | ‖ |
| 32. | ‖ | ‖ | ‖ | ‖ | ‖ |

**33.** Mr. DiCerbo has $200 in his bank account now. If he deposits $10 weekly, which of the following formulas will show his deposits (D) at the end of *n* weeks?

(1) $D = 200 + \frac{10}{n}$
(2) $D = (200)(10n)$
(3) $D = 200 + 10n$
(4) $D = 200 + \frac{n}{10}$
(5) $D = 200n + 10$

**34.** How many cubic yards of concrete are needed to make a cement floor 9′ × 12′ and 6″ thick?
(1) 2 (2) 4 (3) 18 (4) 54 (5) 648

**35.** A new color TV costs $450 in cash. On the installment plan a down payment of $100 plus 18 monthly payments of $22.50 is needed. How much is saved by paying cash?
(1) $5.50 (2) $55 (3) $45 (4) $155 (5) $145

**36.** The tax rate in one school district is $48 per $1,000 of assessed value. What is the tax on a house assessed at $6,600?
(1) $288
(2) $316.80
(3) $336
(4) $416.80
(5) $456.80

**37.** Find the square root of 85 correct to the nearest tenth.
(1) 9.1 (2) 9.2 (3) 9.3 (4) 9.4 (5) 9.5

**38.** Mr. Crandall used 47 gallons of gas on a trip of 611 miles. How many miles per gallon did he get?
(1) 13 (2) 15 (3) 17 (4) 21 (5) 28,717

| | 1 | 2 | 3 | 4 | 5 |
|---|---|---|---|---|---|
| 33. | ‖ | ‖ | ‖ | ‖ | ‖ |
| 34. | ‖ | ‖ | ‖ | ‖ | ‖ |
| 35. | ‖ | ‖ | ‖ | ‖ | ‖ |
| 36. | ‖ | ‖ | ‖ | ‖ | ‖ |
| 37. | ‖ | ‖ | ‖ | ‖ | ‖ |
| 38. | ‖ | ‖ | ‖ | ‖ | ‖ |

39. What is the value of $x$ if:

$$x + 2y = 7$$
$$2x - y = 4$$

(1) 1 (2) 2 (3) 3 (4) $-2$ (5) $-3$

40. If $A \times B = 0$, then:
(1) $A = 0$
(2) $B = 0$
(3) both $A$ and $B$ must be 0
(4) either $A$ or $B$ or both must be 0
(5) both $A$ and $B$ may be unequal to 0

41. An angle is twice as large as its complement. How large is the angle?
(1) 30° (2) 60° (3) 120° (4) 45° (5) 75°

42. Because of a new contract, the hourly wage rate was increased by 6%. If the old rate was $2.50, what is the new rate?
(1) $2.65
(2) $2.85
(3) $3.05
(4) $3.15
(5) $4.00

43. What should be the next term in the series 2, 3, 5, 8, 12, __?
(1) 13 (2) 14 (3) 15 (4) 16 (5) 17

44. The length of a rectangle is 24″ and its width is 7″. How long is a diagonal?
(1) 22″ (2) 25″ (3) 28″ (4) 31″ (5) 34″

45. If $s = \frac{1}{2}gt^2$, find $s$ if $g = 32$ and $t = 3$.
(1) 27 (2) 54 (3) 72 (4) 144 (5) 288

46. If $a$ is a negative number and $ab$ is a positive number, then:
(1) $b$ is positive
(2) $b$ is negative
(3) $a$ is greater than $b$
(4) $b$ is greater than $a$
(5) $b$ may be zero

| | 1 | 2 | 3 | 4 | 5 |
|---|---|---|---|---|---|
| 39. | ‖ | ‖ | ‖ | ‖ | ‖ |
| 40. | ‖ | ‖ | ‖ | ‖ | ‖ |
| 41. | ‖ | ‖ | ‖ | ‖ | ‖ |
| 42. | ‖ | ‖ | ‖ | ‖ | ‖ |
| 43. | ‖ | ‖ | ‖ | ‖ | ‖ |
| 44. | ‖ | ‖ | ‖ | ‖ | ‖ |
| 45. | ‖ | ‖ | ‖ | ‖ | ‖ |
| 46. | ‖ | ‖ | ‖ | ‖ | ‖ |

**47.** An equilateral triangle has the same perimeter as a square whose side is 12″. What is the length of a side of the triangle?
(1) 9″ (2) 12″ (3) 15″ (4) 16″ (5) 18″

**48.** A base angle of an isosceles triangle contains 75°. How large is the vertex angle?
(1) 75° (2) 30° (3) 105° (4) 210° (5) 40°

**49.** MONTHLY SALARY SCALE

| Status | First Year | Second Year | Third Year |
|---|---|---|---|
| Level-1 | $500 | $510 | $515 |
| Level-2 | 520 | 530 | 535 |
| Level-3 | 540 | 550 | 555 |
| Level-4 | 560 | 570 | 575 |
| Level-5 | 580 | 585 | 590 |

Based on the table above, how much more money does a first-year level-5 worker make per month than a third-year level-1 worker?
(1) $15 (2) $165 (3) $65 (4) $80 (5) $75

**50.** HOW THE CONSUMER SPENDS HIS OR HER DOLLAR

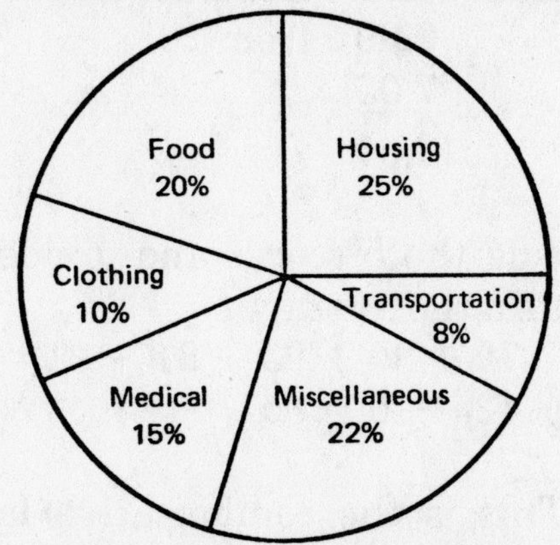


Based on the circle graph above, how much is spent for housing if the annual family income is $10,000?
(1) $2,500
(2) $250
(3) $0.25
(4) $2,200
(5) $25,000

| | 1 | 2 | 3 | 4 | 5 |
|---|---|---|---|---|---|
| 47. | ‖ | ‖ | ‖ | ‖ | ‖ |
| 48. | ‖ | ‖ | ‖ | ‖ | ‖ |
| 49. | ‖ | ‖ | ‖ | ‖ | ‖ |
| 50. | ‖ | ‖ | ‖ | ‖ | ‖ |

# SOLUTIONS

**1.** (5) Be sure to keep the decimal points directly under each other.

$$\begin{array}{r} 2.32 \\ 71.4\phantom{0} \\ \underline{0.003} \\ 73.723 \end{array}$$

**2.** (1)

$$\begin{array}{r} 7527 \\ \underline{-\ 149} \\ 7378 \end{array}$$

**3.** (3) $\frac{5}{6} \times 1\frac{1}{2} = \frac{5}{6} \times \frac{3}{2} = \frac{5}{4}$

**4.** (2)

$$\begin{array}{r} 25.\phantom{5} \\ 3.7\overline{)92.5} \\ \underline{-74\phantom{5}} \\ 185 \\ \underline{-185} \\ 000 \end{array}$$

Move the decimal point one place to the right in both the divisor and the dividend.

**5.** (2) Remember $I = prt$. Interest = principal times rate times time.

$I = (1200)(0.06)(1) = 72.00$

**6.** (4) Add:

$$\begin{array}{r} 3.75 \\ 2.89 \\ \underline{7.05} \\ 13.69 \end{array}$$

Subtract 13.69 from 20:

$$\begin{array}{r} 20.00 \\ \underline{-13.69} \\ 6.31 \end{array}$$

**7.** (3) Add the 5 scores and divide by 5 to obtain the average.

$75 + 85 + 93 + 81 + 91 = 425$

$425 \div 5 = 85$

**8.** (2) This is the commutative law for addition. For multiplication it would be similar:

$3 \times 8 = 8 \times 3$.

**9.** (4)

$$\begin{aligned} 2x + 7 &= 25 \\ 2x &= 18 \\ x &= 9 \end{aligned}$$

(First subtract 7 from each side; then divide each side by 2.)

# SOLUTIONS

Another way to do this problem would be to substitute each of the possible answers in the equation for $x$. The only one to make the equation a true statement is 9.

**10.** (4) First 40 hours at \$2.80 per hour is \$112.00. The overtime is 8 hours at \$4.20 (\$2.80 + \$1.40 for time and a half).

$8 \times 4.20 = 33.60$

$\$112.00 + \$33.60 = \$145.60.$

**11.** (3) All the other fractions are more than $\frac{1}{2}$.

$$\frac{5}{6} > \frac{3}{6} \qquad \frac{2}{3} = \frac{4}{6} > \frac{3}{6}$$

$$\frac{3}{5} = \frac{6}{10} > \frac{5}{10} \qquad \frac{5}{9} = \frac{10}{18} > \frac{9}{18}$$

**12.** (2) 7 *times* 8 = 56. The answer in multiplication is called the *product*.

**13.** (3) To round off 328,637 to the nearest thousand, look at the number in the thousand's place. In this case it is 8. Then look at the number to the right (6). Since it is greater than 5, round off upward to 329,000. One common error would be to neglect the zeros and write the answer as 329.

**14.** (2) Since 1′ = 12″, multiply 12 by 2.54.

$$\begin{array}{r} 2.54 \\ \times\ 12 \\ \hline 508 \\ 254\ \\ \hline 30.48 \end{array}$$

30.48 centimeters

**15.** (5) Change 40% to a decimal, 0.4.

$$0.4 = \frac{x}{30}$$

Multiply both sides by 30.

$0.4(30) = x$

$x = 12.0$

Be careful with the decimal point.

# SOLUTIONS

**16.** (3) Thirteen tenths is the same as $\frac{13}{10}$ or $1\frac{3}{10}$.

$1\frac{3}{10}$ when written as a decimal fraction is 1.3.

**17.** (4) 8 squares out of 16 are shaded. This is $\frac{1}{2}$ of the squares. $\frac{1}{2}$ can be converted to a percent by changing to a decimal (0.50) and moving the decimal point two places to the right.

**18.** (1) To compute the percent of increase, find the amount of increase ($40) and compare this to the original number ($160).

$$\frac{40}{160} = \frac{1}{4} \text{ or } 25\%.$$

**19.** (4) Four tires at $40 would be $160. Now find 20% of $160.

(0.2)(160) = $32 (This is the amount of the reduction.)

$160 − $32 = 128

**20.** (3) Prime factors mean factors which have no divisor except for themselves and one. Factor means multiplier. In each of the other choices there is at least one factor which is not prime.

**21.** (1) Distance = rate × time, or $d = rt$.

Then $t = \frac{d}{r}$ (dividing each side by $r$).

So $t = \frac{200}{r}$.

**22.** (3) Let $n$ be the smaller even number.
Then $n + 2$ is the next consecutive even number.
Sum indicates addition.

$$n + n + 2 = 30$$
$$2n + 2 = 30$$
$$2n = 28$$
$$n = 14$$

# SOLUTIONS

**23.** (1) You don't have to do much work on this one, because 0 times any number is 0.

**24.** (4) First find the common factor.

$12x^2 + 4x - 8 = 4(3x^2 + x - 2)$

Now find the binomial factors of $3x^2 + x - 2$ using the trial and error method.

$3x^2 + x - 2 = (3x \quad )(x \quad )$

$= (3x - 2)(x + 1)$

Thus $12x^2 + 4x - 8 = 4(3x - 2)(x + 1)$.

**25.** (3) If $\frac{1}{2}'' = 1'$, 21′ will have a scale length of $21 \times \frac{1}{2}''$ or $\frac{21}{2}''$, which is the same as $10\frac{1}{2}''$.

**26.** (5) By trying each of the choices for $x$, only 3 works. $5 \times 3 = 15$, which is greater than $2 \times 3 + 7$ or 13.

**27.** (4) 5 times 2′3″ is the same as 10′15″. 15″ = 1′3″, so the correct answer is 11′3″.

**28.** (5) There are 2 pints in a quart and 4 quarts in a gallon. So there are 8 pints in a gallon and 40 pints in 5 gallons. If there are 40 pints in 5 gallons, there are $2 \times 40$ or 80 half-pints in 5 gallons.

**29.** (4) Each of the other choices is the same as $\frac{1}{4}$.

**30.** (4) First subtract 8:35 from 12:00.

12:00 noon is the same as 11:60.

11:60 − 8:35 = 3:25. Now add on 2:15.

3:25 + 2:15 = 5:40.

**31.** (2) In one square yard there are $3 \times 3$ or 9 square feet. Then to change square feet to square yards, *divide* the number of square feet by 9.

$216 \div 9 = 24$ sq. yds.

**32.** (2) An angle greater than 90° but less than 180° is called an *obtuse* angle.

# SOLUTIONS

**33.** (3) Mr. DiCerbo will have \$200 *plus* his weekly deposits. At \$10 per week, at the end of $n$ weeks he will have $10n$ more. The correct answer is $200 + 10n$.

**34.** (1) First change *all* measurements to yards.

$9' = 3$ yds., $12' = 4$ yds., $6'' = \frac{1}{6}$ yd.

Then to find volume, multiply length times width times height.

$3 \times 4 \times \frac{1}{6} = 2$ cubic yards of concrete.

**35.** (2) To find the cost on the installment plan, add the down payment (\$100) to the total monthly charges ($18 \times$ \$22.50).

$100 + 405 =$ \$505 on the installment plan.

\$505 is \$55 more than the cash price of \$450.

**36.** (2) There are 6.6 *thousands* in \$6,600. At 48 for each thousand, the tax is:

$6.6 \times$ \$48 or \$316.80

**37.** (2) Square root means the number which multiplied by itself equals 85 (in this case).

$(9.2)(9.2) = 84.64$, or slightly less than 85.

$(9.3)(9.3) = 86.49$, or slightly more than 85.

9.2 gives the result which is closer to 85.

**38.** (1) In this example, divide the number of miles traveled by the gallons used. This gives you the miles per gallon.

**39.** (3) $x + 2y = 7$ and $2x - y = 4$.

To eliminate $y$ multiply the second equation by 2 and add the two equations.

$$\begin{aligned} x + 2y &= 7 \\ 4x - 2y &= 8 \\ \hline 5x \quad &= 15 \\ x &= 3 \end{aligned}$$

# SOLUTIONS

**40.** (4) If the product of two numbers is zero, one of them (but not necessarily both) must be zero. So choice (4) is the correct choice.

**41.** (2) Complementary angles have a sum of 90°.

Let $2x$ = the angle
$x$ = the complement of the angle
$3x = 90°$
$x = 30°$
$2x = 60°$ is the angle

**42.** (1) 6% of \$2.50 = $0.06 \times 2.50 = 0.15$
The increase is \$0.15. When this is added to \$2.50, the new wage is \$2.65 per hour.

**43.** (5) In the series 2,3,5,8,12 notice the differences. From 2 to 3 is 1, from 3 to 5 is 2, from 5 to 8 is 3 and from 8 to 12 is 4. Then the next number must be 5 more than 12 or 17.

**44.** (2) First draw a diagram.

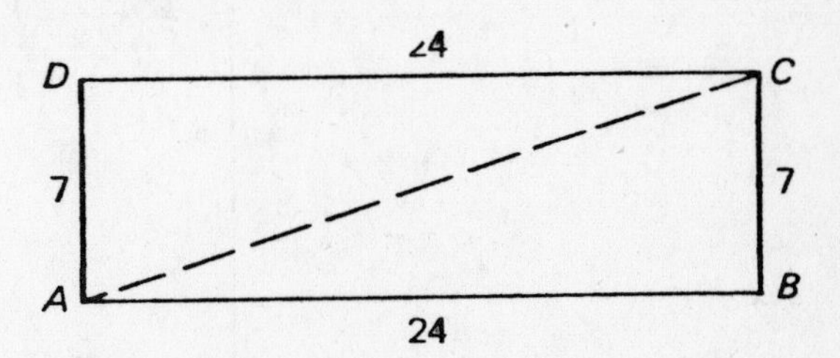


$AC$ is a diagonal ($d$). Use the Pythagorean Theorem, which says, "In a right triangle, $c^2 = a^2 + b^2$."
The hypotenuse ($c$) corresponds with $d$ in the above diagram.

$d^2 = 7^2 + 24^2$
$d^2 = 49 + 576$
$d^2 = 625$

Now find the square root of 625, which is 25.

**45.** (4) Substitute the values for $g$ and $t$.

$$s = \frac{1}{2}gt^2$$

# SOLUTIONS

$$s = \left(\frac{1}{2}\right)(32)(3^2)$$

$$s = 16 \times 9 = 144$$

**46.** (2) Remember that the product of two negative numbers is a positive number, and that the product of a positive number and a negative number is a negative number. Therefore, $b$ must be negative.

**47.** (4) Draw two figures, remembering that perimeter means the sum of the sides.

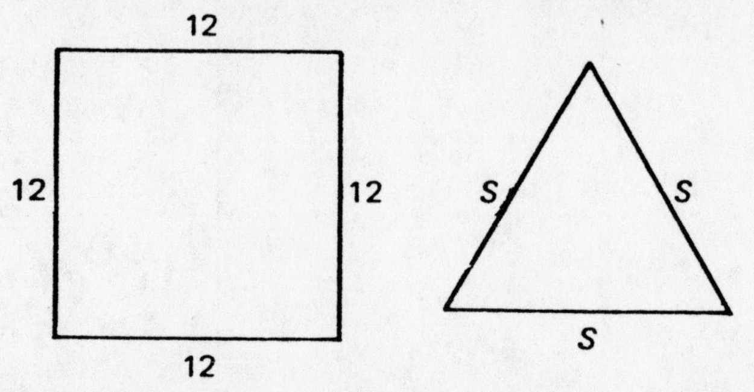


The perimeter of the square is 48. The perimeter of the equilateral triangle must be 48.

$$48 \div 3 = 16$$

**48.** (2) Draw a diagram. Remember that an isosceles triangle has two equal sides and two equal base angles. Remember also that the sum of all the angles of any triangle is 180°.

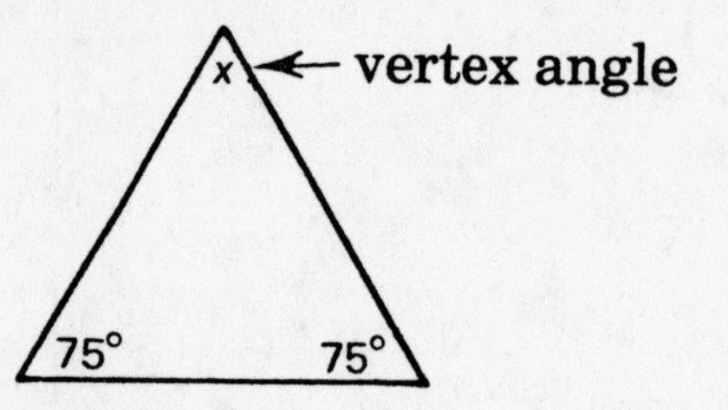


$$x + 75 + 75 = 180°$$
$$x + 150 = 180°$$
$$x = 30° \text{ in the vertex angle.}$$

**49.** (3) The first-year level-5 worker makes $580 per month. The third-year level-1 worker makes $515 per month. The difference is $65.

**50.** (1) 25% of $10,000 is $2,500, the amount spent on housing.

## EVALUATING YOURSELF

To determine the areas in which you may need further study, use the Evaluation Chart below. Notice that the question numbers from the Diagnostic Test appear in the column to the left. Indicate which questions you answered incorrectly. (Unsolved problems are counted as incorrect.) Check the categories of those questions which you did not get right.

For further study in these categories, check the section and page numbers which appear at the right of the chart.

### EVALUATION CHART

| Question | Category | Section | Page |
|---|---|---|---|
| 1 | Decimals | 3.4 | 39 |
| 2 | Whole Numbers | 1.2 | 5 |
| 3 | Fractions | 2.5 | 22 |
| 4 | Decimals | 3.7 | 42 |
| 5 | Percent | 4.4 | 54 |
| 6 | Decimals | 3.4 | 39 |
| 7 | Whole Numbers | 1.1 | 2 |
| 8 | Whole Numbers | 1.1 | 2 |
| 9 | Algebra—Equations | 5.5 | 68 |
| 10 | Decimals | 3.6 | 42 |
| 11 | Fractions | 2.2 | 16 |
| 12 | Whole Numbers | 1.1 | 2 |
| 13 | Whole Numbers | 1.3 | 9 |
| 14 | Decimals | 3.6 | 42 |
| 15 | Percent | 4.1 | 50 |
| 16 | Decimals | 3.1 | 36 |
| 17 | Percent | 4.3 | 53 |
| 18 | Percent | 4.4 | 54 |
| 19 | Percent | 4.4 | 54 |
| 20 | Whole Numbers | 1.2 | 5 |
| 21 | Algebra—Operations | 5.2 | 62 |
| 22 | Algebra—Equations | 5.3 | 64 |
| 23 | Algebra—Evaluation | 5.12 | 77 |
| 24 | Algebra—Factoring | 6.4 | 88 |
| 25 | Geometry—Ratio | 8.6 | 135 |
| 26 | Algebra—Inequality | 7.3 | 104 |
| 27 | Geometry—Measurement | 8.6 | 135 |
| 28 | Fractions | 2.6 | 24 |
| 29 | Fractions—Percent | 4.2 | 50 |
| 30 | Fractions—Time | 2.1 | 14 |
| 31 | Geometry—Area | 8.6 | 135 |
| 32 | Geometry—Angles | 8.1 | 120 |
| 33 | Algebra—Terminology | 5.2 | 62 |
| 34 | Geometry—Solid | 8.8 | 145 |
| 35 | Decimals | 3.6 | 42 |
| 36 | Decimals | 3.7 | 42 |
| 37 | Square Roots | 8.3 | 125 |
| 38 | Whole Numbers | 1.1 | 2 |
| 39 | Algebra—Simultaneous Equations | 7.1 | 100 |

| Question | Category | Section | Page |
|---|---|---|---|
| 40 | Algebra—Quadratic Equations | 7.2 | 102 |
| 41 | Geometry—Angles | 8.1 | 120 |
| 42 | Percent—Decimals | 4.1 | 50 |
| 43 | Number Sequences | 7.5 | 115 |
| 44 | Geometry—Right Triangles | 8.3 | 125 |
| 45 | Algebra—Evaluation | 5.12 | 77 |
| 46 | Algebra—Signed Numbers | 5.4 | 67 |
| 47 | Geometry—Perimeters | 8.6 | 135 |
| 48 | Geometry—Triangles | 8.3 | 125 |
| 49 | Tables | 9.1 | 154 |
| 50 | Graphs—Circle | 9.4 | 159 |

| | |
|---|---|
| TOTAL | 50 |
| (Missed) | |
| SCORE | |

Subtract the number of incorrect or unfinished problems from the total to arrive at your score. If you scored less than 32, you should study the appropriate sections before going on to the Practice Test. Even if you scored 32 or better, it will not hurt to go over the areas in which you still lack skill or confidence.

# Practice Test— General Mathematical Ability

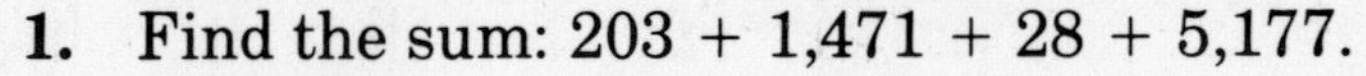

**1.** Find the sum: 203 + 1,471 + 28 + 5,177.
(1) 8,479
(2) 7,131
(3) 7,141
(4) 6,869
(5) 6,879

**2.** Find the quotient of 112,998 divided by 37.
(1) 354 (2) 3,054 (3) 30,054 (4) 353 (5) 3,053

**3.** What is the value of $(0.1)^3$?
(1) 0.3 (2) 0.003 (3) 0.1 (4) 0.001 (5) 3

**4.** Divide $10\frac{1}{2}$ by $1\frac{1}{2}$.
(1) 7 (2) $\frac{1}{7}$ (3) $6\frac{3}{4}$ (4) $\frac{4}{6}$ (5) $6\frac{1}{2}$

**5.** John's new salary is $165 per week. This is a 10% increase over last week's salary. What was his salary last week?
(1) $148.50
(2) $150.00
(3) $151.50
(4) $163.35
(5) $181.50

**6.** Inflation caused the price of a suit to go from $60 to $63. What was the percent of increase in the price?
(1) 3 (2) 4 (3) 5 (4) 6 (5) 8

**7.** A quart of milk and cream contains 28 ounces of milk. If 8 ounces of cream are added, what percent of the new mixture is cream?
(1) 8 (2) 12 (3) 16 (4) 20 (5) 30

| | 1 | 2 | 3 | 4 | 5 |
|---|---|---|---|---|---|
| 1. | ‖ | ‖ | ‖ | ‖ | ‖ |
| 2. | ‖ | ‖ | ‖ | ‖ | ‖ |
| 3. | ‖ | ‖ | ‖ | ‖ | ‖ |
| 4. | ‖ | ‖ | ‖ | ‖ | ‖ |
| 5. | ‖ | ‖ | ‖ | ‖ | ‖ |
| 6. | ‖ | ‖ | ‖ | ‖ | ‖ |
| 7. | ‖ | ‖ | ‖ | ‖ | ‖ |

**8.** Mr. Amell insured his house for $14,500. If the rate per year is $0.44 per hundred, find his annual premium.
(1) $53.80
(2) $538.00
(3) $63.80
(4) $638
(5) $44

| | 1 | 2 | 3 | 4 | 5 |
|---|---|---|---|---|---|
| 8. | ‖ | ‖ | ‖ | ‖ | ‖ |
| 9. | ‖ | ‖ | ‖ | ‖ | ‖ |
| 10 | ‖ | ‖ | ‖ | ‖ | ‖ |
| 11. | ‖ | ‖ | ‖ | ‖ | ‖ |
| 12. | ‖ | ‖ | ‖ | ‖ | ‖ |

**9.** Mr. Nolan started the month with a balance of $480.70 in his checking account. If he drew checks for $23.80, $65 and $165, and made a deposit of $200, what was his new balance?
(1) $426.90
(2) $26.90
(3) $226.90
(4) $427.90
(5) $27.90

**10.** A refrigerator can be purchased for $400 cash, or 25% down with 12 monthly installments of $30. How much is saved by paying cash?
(1) $30 (2) $40 (3) $60 (4) $80 (5) $100

**11.** The distance around a tree is 44″. The diameter of the tree is approximately how many inches?
(1) 7 (2) 14 (3) 21 (4) 22 (5) 24

**12.** The perimeter of a rectangle is 38″. If the length is 3″ more than the width, find the width.
(1) $17\frac{1}{2}''$ (2) 11″ (3) $14\frac{1}{2}''$ (4) 16″ (5) 8″

**13.**

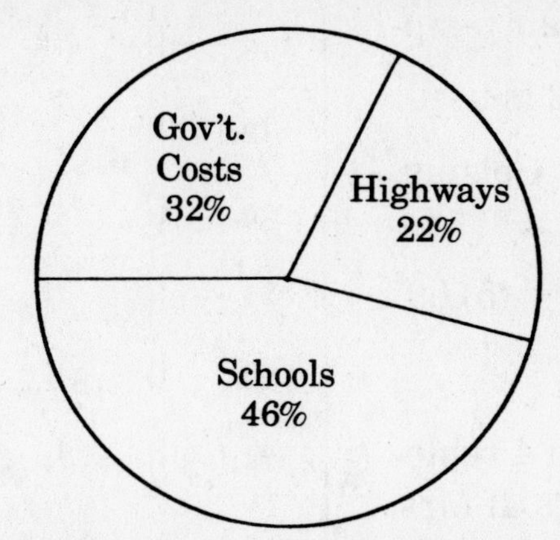


The circle graph illustrates how local tax money is spent. If Mr. Mead paid a total local tax of $640, how much went for schools?
(1) $294.40
(2) $140.80
(3) $2,944
(4) $9.44
(5) $14.08

**14.** What number belongs next in the series 7,9,15,17,23,__?
(1) 24 (2) 25 (3) 29 (4) 31 (5) 33

**15.** How many inches are contained in $f$ feet and $i$ inches?
(1) $f \times i$
(2) $f + i$
(3) $f + 12i$
(4) $12f + i$
(5) $\frac{f}{12} + i$

**16.** Solve the following equation for $x$:

$$2x + 7 = 31$$

(1) 11 (2) 12 (3) 13 (4) $16\frac{1}{2}$ (5) 19

**17.** If the numerator and denominator of a fraction are each multiplied by 2, then the value of the fraction is:
(1) unchanged
(2) multiplied by 2
(3) divided by 2
(4) multiplied by 4
(5) multiplied by $\frac{1}{2}$

| | 1 | 2 | 3 | 4 | 5 |
|---|---|---|---|---|---|
| 13. | ‖ | ‖ | ‖ | ‖ | ‖ |
| 14. | ‖ | ‖ | ‖ | ‖ | ‖ |
| 15. | ‖ | ‖ | ‖ | ‖ | ‖ |
| 16. | ‖ | ‖ | ‖ | ‖ | ‖ |
| 17. | ‖ | ‖ | ‖ | ‖ | ‖ |

**18.** To change from Celsius to Fahrenheit temperature the formula $F = (\frac{9}{5})(C) + 32$ is used. When the temperature is 20° Celsius, what is it on the Fahrenheit scale?

(1) $93\frac{3}{5}°$ (2) 78° (3) 58° (4) $62\frac{3}{5}°$ (5) 68°

**19.** A recipe calls for 6 cups of flour and 4 tablespoons of butter, among other things. If Mrs. Boynton has only 4 cups of flour, how many tablespoons of butter should she use?

(1) 2 (2) 3 (3) 4 (4) $2\frac{1}{3}$ (5) $2\frac{2}{3}$

**20.** Find the product of $(a + 2b)$ and $(a - 2b)$.
(1) $a^2 - 4b^2$
(2) $2a$
(3) $-4b$
(4) $a^2 - 4ab - 4b^2$
(5) $a^2 + 4ab - 4b^2$

**21.** If $x^2 = 40$, what is the positive value of $x$ correct to the nearest tenth?
(1) 20 (2) 6.3 (3) 6.4 (4) 6.5 (5) 6.6

**22.** Mr. Holman is a salesman who receives a weekly salary of $50 plus a commission of 8% on all his sales. One week he had $1,650 worth of sales. What was his total salary?
(1) $182
(2) $132
(3) $216
(4) $1,320
(5) $1,370

**23.** A 15′ long ladder leans against a house. If the foot of the ladder is 9′ from the house, how high up the house will the ladder reach?
(1) 6′ (2) 8′ (3) 9′ (4) 10′ (5) 12′

| | 1 | 2 | 3 | 4 | 5 |
|---|---|---|---|---|---|
| 18. | ‖ | ‖ | ‖ | ‖ | ‖ |
| 19. | ‖ | ‖ | ‖ | ‖ | ‖ |
| 20. | ‖ | ‖ | ‖ | ‖ | ‖ |
| 21. | ‖ | ‖ | ‖ | ‖ | ‖ |
| 22. | ‖ | ‖ | ‖ | ‖ | ‖ |
| 23. | ‖ | ‖ | ‖ | ‖ | ‖ |

**24.**

| x | 1 | 2 | 3 | 4 |
|---|---|---|---|---|
| y | 3 | 5 | 7 | 9 |

The table above expresses a relationship between $x$ and $y$. Which of the following is a correct algebraic statement of that relationship?
(1) $y = x + 2$
(2) $y = 3x$
(3) $y = 2x$
(4) $y = 2x + 1$
(5) $x = 2y + 1$

**25.** The ratio of 4″ to 2 yards is the same as which of the following fractions?
(1) 2 : 2 (2) 2 : 4 (3) 1 : 18 (4) 18 : 1 (5) 1 : 6

**26.** In a certain store, steak costs $1.85 a pound. How much change from a $10 bill will Mrs. Pirri get if she buys a 3 pound steak?
(1) $5.45 (2) $5.55 (3) $4.45 (4) $4.55 (5) $5.65

**27.** Evaluate the following expression if $a = 1$, $b = 2$, $c = -1$:

$$8ca^2 + 4b$$

(1) 0 (2) 8 (3) −8 (4) 4 (5) −4

**28.**

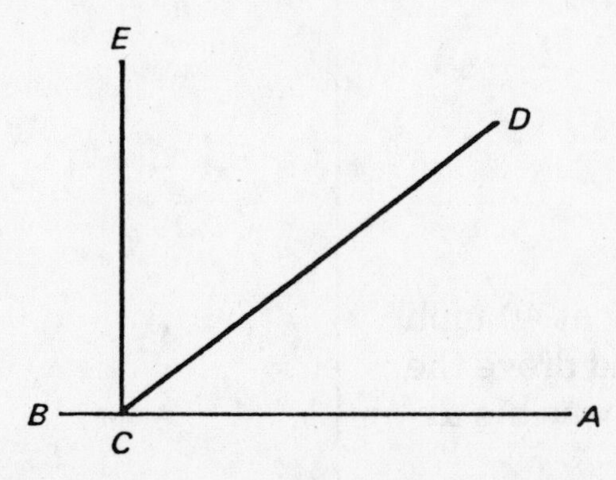


In the above drawing $AB$ is perpendicular to $EC$. $\sphericalangle ECD$ is twice the size of $\sphericalangle ACD$. How many degrees are in $\sphericalangle ACD$?
(1) 30 (2) 60 (3) 45 (4) 120 (5) 25

| | 1 | 2 | 3 | 4 | 5 |
|---|---|---|---|---|---|
| 24. | ‖ | ‖ | ‖ | ‖ | ‖ |
| 25. | ‖ | ‖ | ‖ | ‖ | ‖ |
| 26. | ‖ | ‖ | ‖ | ‖ | ‖ |
| 27. | ‖ | ‖ | ‖ | ‖ | ‖ |
| 28. | ‖ | ‖ | ‖ | ‖ | ‖ |

29.

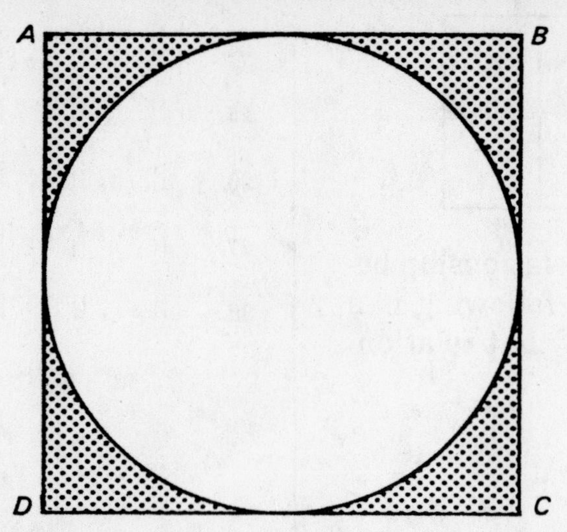


*ABCD* is a square 14″ on a side. Approximately how many square inches are contained in the shaded area?

(1) 42 (2) 154 (3) 196 (4) 174 (5) 152

30. A blueprint is drawn to the scale $\frac{1}{2}''$ = 1′. How long is a room $10\frac{1}{2}''$ long in the blueprint?

(1) 20′ (2) $10\frac{1}{2}'$ (3) $10\frac{1}{2}''$ (4) 21′ (5) 22′

31. Solve the following equation for $x$:

$$x^2 = 2x + 15$$

(1) −3
(2) 5
(3) {3,−5}
(4) {5,−3}
(5) {7,−3}

32. Mr. Larson drove his car steadily at 40 mph for 120 miles. He then sped up and drove the next 120 miles at 60 mph. What was his average speed?

(1) 50 mph
(2) 48 mph
(3) 52 mph
(4) 46 mph
(5) 54 mph

| | 1 | 2 | 3 | 4 | 5 |
|---|---|---|---|---|---|
| 29. | ‖ | ‖ | ‖ | ‖ | ‖ |
| 30. | ‖ | ‖ | ‖ | ‖ | ‖ |
| 31. | ‖ | ‖ | ‖ | ‖ | ‖ |
| 32. | ‖ | ‖ | ‖ | ‖ | ‖ |

**33.** Solve the following equations for $x$:

$$5x + 4y = 27$$
$$x - 2y = 11$$

(1) 6 (2) 3 (3) 4 (3) −7 (5) 7

**34.** One of two supplementary angles is 9 times as large as the other. How large is the smaller of the two angles?
(1) 9° (2) 10° (3) 18° (4) 20° (5) 36°

**35.** Find the smallest angle of a triangle if the second is 3 times the first and the third is 5 times the first.
(1) 18° (2) 20° (3) 9° (4) 10° (5) 40°

**36.** How much money must be invested at 6% interest to yield an income of $216 per year?
(1) $12.96
(2) $36
(3) $360
(4) $3,600
(5) $36,000

**37.**

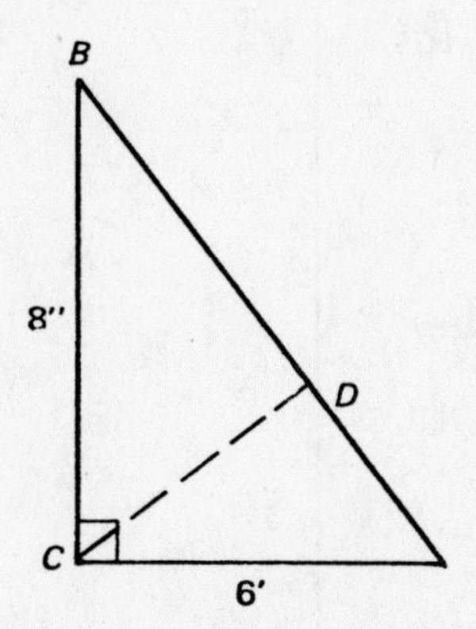


In the above drawing, ∢ $C$ is a right angle $CD$ is perpendicular to $AB$. How long is $CD$?
(1) 10″ (2) 6″ (3) 8″ (4) 4.8″ (5) 5.2′

**38.** Mrs. Streeter's hourly wage was $2.80. Recently she received a 25% increase. What is her new hourly rate?
(1) $2.10
(2) $3.50
(3) $3.05
(4) $2.55
(5) $3.16

| | 1 | 2 | 3 | 4 | 5 |
|---|---|---|---|---|---|
| 33. | ‖ | ‖ | ‖ | ‖ | ‖ |
| 34. | ‖ | ‖ | ‖ | ‖ | ‖ |
| 35. | ‖ | ‖ | ‖ | ‖ | ‖ |
| 36. | ‖ | ‖ | ‖ | ‖ | ‖ |
| 37. | ‖ | ‖ | ‖ | ‖ | ‖ |
| 38. | ‖ | ‖ | ‖ | ‖ | ‖ |

**39.** Grass seed costs $1.10 a pound and will cover 500 sq. ft. of lawn. How much will seed cost for a lawn 100′ by 200′?
(1) $66 (2) $4.40 (3) $44 (4) $440 (5) $22

**40.** Small patio blocks are 6″ by 12″. How many will be needed for a 9′ by 12′ patio?
(1) 108 (2) 54 (3) 162 (4) 216 (5) 72

**41.** Solve the following equation for $x$:

$$x^2 + 3x - 28 = 0$$

(1) 4 (2) −7 (3) {4,7} (4) {−4,7} (5) {4,−7}

**42.** If $a$ is a positive number and $b$ is a negative number, which of the following must be a negative number?
(1) $-ab$
(2) $a^2 - b^2$
(3) $a + 2b$
(4) $a - b$
(5) $ab$

**43.** Which of the given values is a solution for $3x + 2 < 7$?
(1) 1 (2) 2 (3) 3 (4) 4 (5) 5

**44.** Our dog eats $\frac{3}{4}$ of a can of dog food each day. If dog food is priced at 2 cans for 47 cents, how much will food for 16 days cost?
(1) $1.98
(2) $7.52
(3) $1.41
(4) $2.82
(5) $5.64

**45.** During 1972, the Mets drew about 2,028,000 in 78 home games. What was their average attendance per game?
(1) 26,000
(2) 28,000
(3) 3,000
(4) 32,000
(5) 2,600

| | 1 | 2 | 3 | 4 | 5 |
|---|---|---|---|---|---|
| 39. | ‖ | ‖ | ‖ | ‖ | ‖ |
| 40. | ‖ | ‖ | ‖ | ‖ | ‖ |
| 41. | ‖ | ‖ | ‖ | ‖ | ‖ |
| 42. | ‖ | ‖ | ‖ | ‖ | ‖ |
| 43. | ‖ | ‖ | ‖ | ‖ | ‖ |
| 44. | ‖ | ‖ | ‖ | ‖ | ‖ |
| 45. | ‖ | ‖ | ‖ | ‖ | ‖ |

**46.** A color TV listing at $540 was reduced in price by 20%. After a sales tax of 7%, what was the cost of the TV set?
(1) $507.24
(2) $432
(3) $408
(4) $474.48
(5) $462.24

**47.** The Yankees won 55% of their first hundred games. How many of their next 50 games must they win to be at or over .600? (Over .600 means winning 60% of their games.)
(1) 40 (2) 45 (3) 50 (4) 35 (5) 30

**48.** A good rule of thumb is that a house should cost no more than $2\frac{1}{2}$ times its owner's income. How much should you be earning to afford a $28,000 home?
(1) $12,000
(2) $10,000
(3) $11,200
(4) $13,200
(5) $14,000

**49.**

STANDINGS

| | W | L |
|---|---|---|
| Los Angeles | 24 | 26 |
| New York | 21 | 29 |
| Cincinnati | 19 | 31 |
| Miami | 40 | 10 |
| Philadelphia | 35 | 15 |

(*W* means number of games won; *L* means number of games lost.)

What percent of its games did Miami win?
(1) 0.80% (2) 80% (3) 40% (4) 50% (5) 30%

| | 1 | 2 | 3 | 4 | 5 |
|---|---|---|---|---|---|
| 46. | ‖ | ‖ | ‖ | ‖ | ‖ |
| 47. | ‖ | ‖ | ‖ | ‖ | ‖ |
| 48. | ‖ | ‖ | ‖ | ‖ | ‖ |
| 49. | ‖ | ‖ | ‖ | ‖ | ‖ |

**50.** MILES OF STATE HIGHWAYS

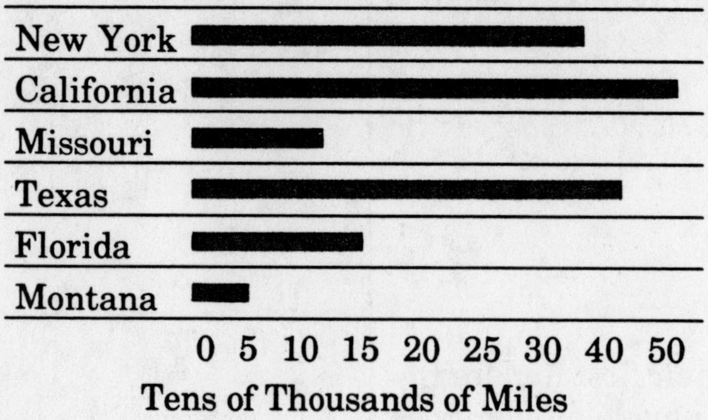

New York
California
Missouri
Texas
Florida
Montana

0 5 10 15 20 25 30 40 50

Tens of Thousands of Miles

California has how many times as many miles of state highway as Montana?

(1) 100 (2) 45 (3) 10 (4) 35 (5) 55

50. 1 2 3 4 5

# SOLUTIONS

1. (5) Be careful to line up the numbers accurately.

2. (2) A common error is to leave out the *zero* as a place holder.

$$\begin{array}{r} 3054 \\ 37\overline{)112998} \\ -\underline{111\phantom{98}} \\ 199\phantom{8} \\ -\underline{185\phantom{8}} \\ 148 \\ -\underline{148} \end{array}$$

3. (4) $(0.1)^3$ means use 0.1 as a multiplier 3 times.

   $(0.1)^3 = (0.1)(0.1)(0.1) = 0.001$

   REMEMBER: when multiplying decimals, count off one place in the answer for each decimal place.

4. (1) As usual, change each mixed number to an improper fraction and invert the divisor.

$$10\frac{1}{2} \div 1\frac{1}{2} = \frac{21}{2} \div \frac{3}{2} = \frac{21}{2} \times \frac{2}{3} = 7$$

5. (2) \$165 must be 110% of the old salary. Use $x$ as the old salary.

   110% of $x = 165$ (110% = 1.1)

   $1.1x = 165$ (Now divide each side by 1.1.)

   $x = \$150$

6. (3) The increase was \$3. This must be compared to the original cost. $\frac{3}{60} = \frac{1}{20}$.

   Convert $\frac{1}{20}$ to a decimal: 0.05.

   0.05 written as a percent is 5%.

7. (5) A quart has 32 ounces. Since 28 ounces are milk, 4 ounces are cream. If 8 ounces of cream are added, there is now a total of 40 ounces in the mixture, with 12 ounces of cream. Compare 12 to 40 and

# SOLUTIONS

convert to a percent.

$$\frac{12}{40} = \frac{3}{10} = 0.30 = 30\%$$

**8.** (3) There are 145 *hundreds* in $14,500. (The 5 is in the hundreds place.) Since the rate is $0.44 per hundred, multiply:
(0.44)(145) = $63.80.

**9.** (1) The easy way to do this problem is to add the balance and the deposits:
$480.70 + $200 = $680.70
Now add the checks drawn:
$23.80 + $65 + $165 = $253.80
Find the difference:
680.70 − 253.80 = $426.90 (the balance at the end of the month).

**10.** (3) Paying on time, you need 25% of $400 or $100 as a down payment. Twelve installments of $30 each is 12 × $30 or $360.
$100 + 360 = $460, which is $60 more than the cash price.

**11.** (2) The circumference of a circle (the tree) is given by the formula $C = \pi d$, where $C$ is the circumference, $\pi$ is about $\frac{22}{7}$ and $d$ is the diameter.

$$C = \pi d$$

$$44 = \frac{22}{7} \times d$$

To solve, multiply both sides by $\frac{7}{22}$.

$$\frac{7}{22} \times 44 = \frac{7}{22} \times \frac{22}{7} \times d$$

$$14 = d$$

**12.** (5) Perimeter means the sum of the sides. Draw a diagram, and label it to attack the problem.

# SOLUTIONS

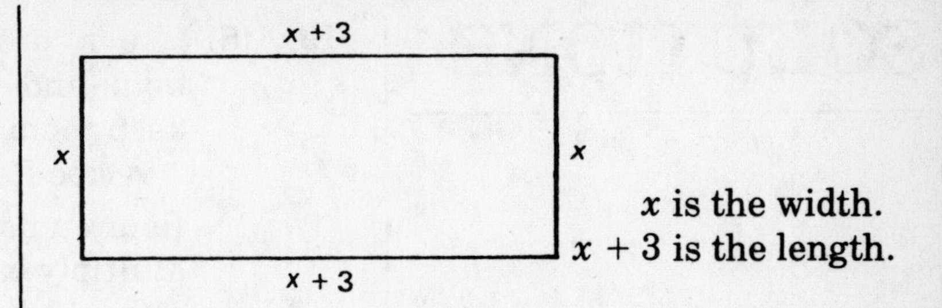


$x$ is the width.
$x + 3$ is the length.

The perimeter is $x + x + 3 + x + x + 3$, or $4x + 6$.

$$4x + 6 = 38$$
$$4x = 32$$
$$x = 8 \text{ for the width}$$

**13.** (1) Since schools account for 46% of the budget, find 46% of \$640.

$(0.46)(640) = \$294.40$

**14.** (2) If you look at the numbers in pairs, 9 is 2 more than 7, 17 is 2 more than 15. This clue suggests that the next number is 2 more than 23, or 25.

**15.** (4) In 1′ there are 12″; in 2′ there are 24″ (or $12 \times 2$). Thus in $f$ feet, there must be $12 \times f$ inches. Now add on the $i$ inches to obtain a total of $12f + i$.

**16.** (2) $2x + 7 = 31$

$$2x = 31 - 7 = 24$$
$$\frac{2x}{2} = \frac{24}{2}$$
$$x = 12$$

**17.** (1) If the numerator and denominator are multiplied by the same number (except 0), the effect is to multiply by 1, leaving the fraction unchanged.

**18.** (5) Use $F = (\frac{9}{5})(C) + 32$

Substitute 20° for $C$.

$$F = (\frac{9}{5})(20) + 32$$
$$F = 36 + 32 = 68°$$

# SOLUTIONS

**19.** (5) Use a proportion, comparing flour to flour and butter to butter. Be sure to keep them in the proper order.

$$\begin{array}{r} \text{recipe flour} \rightarrow \\ \text{flour on hand} \rightarrow \end{array} \frac{6}{4} = \frac{4}{x} \begin{array}{l} \leftarrow \text{recipe butter} \\ \leftarrow \text{butter needed} \end{array}$$

Multiply each side by $4x$:

$$6x = 16$$

$$x = \frac{16}{6}$$

$$x = \frac{8}{3} \text{ or } 2\frac{2}{3} \text{ tablespoons.}$$

**20.** (1)

$$\begin{array}{rr} & a - 2b \\ \times & a + 2b \\ \hline & 2ab - 4b^2 \\ a^2 - 2ab & \\ \hline a^2 & - 4b^2 \end{array}$$

**21.** (2) The safest way to handle this question is to square each answer, and find out which is closest to 40.

$(6.3)^2 = (6.3)(6.3)$ or 39.69

$(6.4)^2 = (6.4)(6.4)$ or 40.96

The first is too small by 0.31 and the second is too large by 0.96. Therefore, 6.3 is the better choice.

**22.** (1) Add the salary, \$50, to the commission, (0.08)(1,650) or \$132.

\$50 + \$132 = \$182.

**23.** (5) A diagram is necessary.

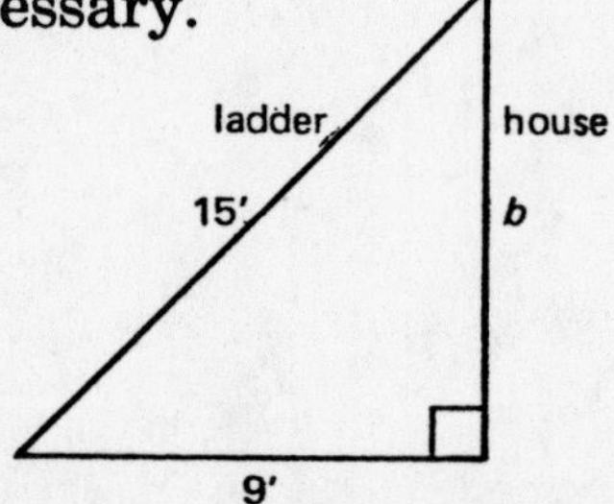


This is an application of the Pythagorean Theorem.

In a right triangle, $c^2 = a^2 + b^2$.

$c = 15, a = 9, b = ?$

$225 = 81 + b^2$

$144 = b^2$

$b = 12$

# SOLUTIONS

**24.** (4) In this case, it's best to try each answer. But be careful not to base your conclusion on only one case. Be sure that all cases satisfy the equation you choose.

**25** (3) Two yards is $2 \times 36''$ or $72''$. The ratio of $4''$ to $72''$ is $\frac{4}{72}$ or $\frac{1}{18}$. Remember that in a ratio, both numerator and denominator must be in the same units.

**26.** (3) Multiply 3 times \$1.85 to get \$5.55 for the cost of the steak. Subtract from \$10.00 to get \$4.45.

**27.** (1) $a = 1, b = 2, c = -1$

$$\begin{aligned} 8ca^2 + 4b &= (8)(-1)(1)^2 + (4)(2) \\ &= (8)(-1)(1) + 8 \\ &= -8 + 8 \\ &= 0 \end{aligned}$$

**28.** (1) Since $AB \perp EC$, $\sphericalangle\ ACE$ is a right angle containing 90°.

$$\begin{aligned} \text{Let } \sphericalangle\ ACD &= x^\circ \\ \text{and } \sphericalangle\ ECD &= 2x^\circ \\ \text{Then } x + 2x &= 90^\circ \\ 3x &= 90^\circ \\ x &= 30^\circ \end{aligned}$$

**29.** (1) To find the shaded area, you must subtract the area of the circle from that of the square.

$A = s^2$ is the area of a square.

$A = (14)^2$ or 196 sq. in. in the square.

The radius of the circle must be half the side of the square, or 7″.

$A = \pi r^2$ is the area of a circle.

$A = \left(\frac{22}{7}\right)(7)(7)$ or 154 sq. in. in the circle.

$196 - 154 = 42$ sq. in. shaded.

**30.** (4) This question asks, "How many halves are there in $10\frac{1}{2}$?"

# SOLUTIONS

$$10\frac{1}{2} \div \frac{1}{2} = \frac{21}{2} \div \frac{1}{2}$$
$$= \frac{21}{2} \times \frac{2}{1}$$
$$= 21$$

The room is 21′ long.

**31.** (4) First, recognize this as a quadratic equation. Arrange all terms on the left side in descending order of exponents.

$$x^2 = 2x + 15$$
$$x^2 - 2x - 15 = 0$$

Factor the left side:

$$(x + 3)(x - 5) = 0$$

Set each factor equal to zero:

$$x + 3 = 0 \qquad x - 5 = 0$$
$$x = -3 \qquad x = 5$$

$\{5,-3\}$ is the solution.

**32.** (2) Be careful. This is a trick question. The answer is *not* 50 mph. Mr. Larson used 3 hours for the first part of his trip $(\frac{120}{40})$ and 2 hours for the second part $(\frac{120}{60})$. He used 5 hours for 240 miles, so his average speed was $240 \div 5$ or 48 mph.

**33.** (5) $5x + 4y = 27$
$x - 2y = 11$

To eliminate $y$, multiply both sides of the second equation by 2 and add it to the first equation.

$$5x + 4y = 27$$
$$2x - 4y = 22$$
$$7x = 49$$
$$x = 7$$

**34.** (3) Supplementary means they add up to 180°.

Call the angles $x$ and $9x$.

$$x + 9x = 180°$$
$$10x = 180°$$
$$x = 18°$$

# SOLUTIONS

**35.** (2) The angles of a triangle have a sum of 180°.

$$x + 3x + 5x = 180°$$
$$9x = 180°$$
$$x = 20°$$

**36.** (4) Think "6% of some number is $216."
Call the number $x$.

$0.06x = 216$

Divide each side by 0.06:

$$0.06\overline{)216.00} = 3600.$$

**37.** (4) First find $AB$ by the Pythagorean Theorem.

$(AB)^2 = 6^2 + 8^2$
$(AB)^2 = 36 + 64 = 100$
$AB = 10$

Now, the area of a triangle is
$A = \frac{1}{2} \times b \times h.$

In this case $A = \frac{1}{2} \times 6 \times 8$ or 24.

But, you could call $AB$ the base and $CD$ the height.
Since $AB = 10$, using $A = \frac{1}{2}bh$,

$24 = \frac{1}{2} \times 10 \times h$
$24 = 5h$
$h = \frac{24}{5} = 4.8$

**38.** (2) To find the new hourly rate, first find 25% or $\frac{1}{4}$ of \$2.80.

$\frac{1}{4} \times 2.80 = 0.70$

\$0.70 is the increase.

\$2.80 + \$0.70 = \$3.50

**39.** (3) A 100′ by 200′ lot has 20,000 sq. ft. To find the pounds needed, divide 20,000 by 500.

20,000 ÷ 500 = 40 lbs. required.
40 × \$1 10 = \$44.

# SOLUTIONS

**40.** (4) A 6″ by 12″ block is exactly $\frac{1}{2}$ of one sq. ft.

(6″ = $\frac{1}{2}$′, 12″ = 1′ and $\frac{1}{2} \times 1 = \frac{1}{2}$). The patio will contain $9 \times 12$ or 108 sq. ft. How many halves are in 108?

$108 \div \frac{1}{2} = 108 \times \frac{2}{1} = 216$ blocks needed.

**41.** (5) First factor $x^2 + 3x - 28$.

$(x - 4)(x + 7)$

Set each of the two factors equal to zero and solve.

$x - 4 = 0 \qquad x + 7 = 0$

$x = 4 \qquad x = -7$

**42.** (5) Since $a$ is positive and $b$ is negative, only $ab$ *must* be negative. The product of a positive number and a negative number is negative.

**43.** (1) Trying each choice, we find that only (1) works. Remember that the symbol $<$ means "less than."

**44.** (4) 16 days at $\frac{3}{4}$ can per day means 12 cans will be needed.

12 is 6 groups of 2 cans and

$6 \times \$0.47 = \$2.82$.

**45.** (1) Just divide 2,028,000 by 78.

```
        26,000
78 /2,028,000
     1 56
      468
      468
        0
```

**46.** (5) First take 20% of $540.

$0.20 \times \$540 = \$108$

Reduce $540 by 108 to get the sale price of $432.

Now add on the tax of 7%.

$(0.07)(432) = \$30.24$

$\$432 + \$30.24 = \$462.24$

# SOLUTIONS

**47.** (4) 55% of 100 is 55 games won in the first hundred. 60% of 150 (100 + 50) is 90. A total of 90 games must be won to achieve .600.

$90 - 55 = 35$

The team must win 35 more games.

**48.** (3) This time, use $x$ as the owner's income.

$$2\frac{1}{2}x = \$28{,}000$$

$$x = 28{,}000 \div 2\frac{1}{2}$$

$$x = 28{,}000 \times \frac{2}{5} = \$11{,}200$$

$11,200 is the minimum salary you should earn to afford the house.

**49.** (2) Miami played a total of 50 games. It won $\frac{40}{50}$ of the games played. Change $\frac{40}{50}$ to percent by dividing 50 into 40. The answer is 0.80. To convert to percent, move the decimal point two places to the right. The correct answer is 80%.

**50.** (3) California has 500,000 miles of state highway and Montana has 50,000. California has 10 times as much state highway as Montana.

## EVALUATING YOURSELF

To determine the areas in which you may need further study, use the Evaluation Chart below. Notice that the question numbers from the Practice Test appear in the column to the left. Indicate which questions you answered incorrectly. (Unsolved problems are counted as incorrect.) Check the categories of those questions which you did not get right.

For further study in these categories, check the section and page numbers which appear at the right of the chart.

### EVALUATION CHART

| Question | Category | Section | Page |
|---|---|---|---|
| 1 | Whole Numbers | 1.1 | 2 |
| 2 | Whole Numbers | 1.2 | 5 |
| 3 | Exponents | 1.2 | 5 |
| 4 | Fractions | 2.6 | 24 |
| 5 | Percent | 4.4 | 54 |
| 6 | Percent | 4.4 | 54 |
| 7 | Percent | 4.4 | 54 |
| 8 | Decimals | 3.6 | 42 |
| 9 | Decimals | 3.4 | 39 |
| 10 | Percent | 4.4 | 54 |
| 11 | Geometry—Circles | 8.7 | 140 |
| 12 | Geometry—Perimeter | 8.6 | 135 |
| 13 | Circle Graphs | 9.4 | 159 |
| 14 | Number Sequences | 7.5 | 115 |
| 15 | Algebra—Operations | 5.2 | 62 |
| 16 | Algebra—Equations | 5.5 | 68 |
| 17 | Fractions | 2.3 | 19 |
| 18 | Algebra—Evaluation | 5.12 | 77 |
| 19 | Fractions—Ratio | 5.1 | 62 |
| 20 | Algebra | 5.8 | 74 |
| 21 | Square Roots | 8.3 | 125 |
| 22 | Percent | 4.4 | 54 |
| 23 | Geometry—Area | 8.6 | 135 |
| 24 | Algebra—Linear Equations | 7.4 | 108 |
| 25 | Fractions | 2.2 | 17 |
| 26 | Decimals | 3.5 | 41 |
| 27 | Algebra | 5.12 | 77 |
| 28 | Geometry—Angles | 8.1 | 120 |
| 29 | Geometry—Area | 8.6 | 135 |
| 30 | Fractions—Ratio | 2.2 | 16 |
| 31 | Algebra—Quadratic Equations | 7.2 | 102 |
| 32 | Whole Numbers | 1.2 | 5 |
| 33 | Algebra—Simultaneous Equations | 7.1 | 100 |
| 34 | Geometry—Angles | 8.1 | 120 |
| 35 | Geometry—Triangles | 8.3 | 125 |
| 36 | Percent | 4.4 | 54 |
| 37 | Geometry—Triangles | 8.5 | 132 |
| 38 | Percent | 4.4 | 54 |
| 39 | Geometry—Area | 8.6 | 135 |

| Question | Category | Section | Page |
|---|---|---|---|
| 40 | Geometry—Area | 8.6 | 135 |
| 41 | Algebra—Quadratic Equations | 7.2 | 102 |
| 42 | Algebra—Signed Numbers | 5.4 | 67 |
| 43 | Algebra—Inequalities | 7.3 | 104 |
| 44 | Fractions | 2.6 | 24 |
| 45 | Whole Numbers | 1.2 | 5 |
| 46 | Percent | 4.4 | 54 |
| 47 | Percent | 4.4 | 54 |
| 48 | Fractions | 2.5 | 22 |
| 49 | Tables | 9.1 | 154 |
| 50 | Bar Graphs | 9.2 | 156 |

| | |
|---|---|
| TOTAL | 50 |
| (Missed) | |
| SCORE | |

Subtract the number of incorrect or unfinished problems from the total to arrive at your score.

# INDEX